Workbook

Modern Carpentry

12th Edition

by

Willis H. Wagner

Howard Bud Smith

Mark W. Huth

Publisher
The Goodheart-Willcox Company, Inc.
Tinley Park, IL
www.g-w.com

Introduction

This workbook has been prepared for use with the textbook, *Modern Carpentry*. The workbook is designed to help you in your study of carpentry. It will reinforce your understanding of the correct procedures and the vast amount of technical information connected with the trade.

The chapters in the workbook correlate with those in the textbook. The order of the individual questions and problems also follow the same sequence as the textbook material. This will make it easier for you to find information in the textbook when checking your answers.

To make this workbook a more effective guide to learning, we suggest that you first study the assigned material in the textbook. As you study, pay close attention to both the text and the illustrations. The drawings and photographs contain a great deal of important information. Study each illustration and its caption carefully until you understand what it is conveying.

After you have completed the study assignment, lay the textbook aside while you answer the problems in the workbook. When filling in words, be sure they are spelled correctly.

As you work your way through each chapter, fill in the blanks whenever you are relatively certain of the answer. Guessing is not really helpful in a learning activity. Use the text to find the answers for the questions that you missed. Check for correctness of your answers. Do not just copy the answer from the book. Be certain that you understand the principles, practices, or information relating to the answer.

A number of the chapters in the workbook include problems in estimating amounts of materials and other calculations. Work out the problem in the space provided so that your instructor can check your figures and determine if you have followed an efficient procedure. It will be best to list the steps in your calculations as shown in the textbook. Following this procedure will not only make it easier for your instructor to check your work, but will also keep you from becoming "lost" in complicated problems.

Table of Contents

Section 5—Special Construction

Section 6—Mechanical Systems

Name _Matthew Horner_   Date _08/21/18_   Score _____

CHAPTER **1**

The Carpenter's Workplace

Carefully study the chapter and then answer the following questions.

Increase 1. The number of carpenters employed in the United States is expected to ____ over the next several years.
    A. decrease
    B. remain about the same
    C. increase

More technical 2. Carpentry jobs will become ____ in years to come.
    A. more technical
    B. more strenuous
    C. less desirable
    D. All of the above.

Heating, Plumbing, Electrical 3. Since skilled carpenters (journeymen) are capable of performing a variety of jobs on a regular construction site, they can usually fill the requirements for a broad range of related work. From the following list, select the task that a skilled carpenter (journeyman) would _not_ likely be able to handle without additional training.
    A. Drywall work.
    B. Insulation.
    C. Heating, plumbing, and electrical maintenance in commercial buildings.
    D. Roofing.

four 4. Approximately _4_ out of every 10 carpenters is/are self-employed.

about the same 5. Compared to other trades, carpenters' income is ____.
    A. about the same
    B. significantly higher
    C. significantly lower

All of the above 6. Students enrolled in vocational-technical schools who are interested in becoming carpenters should take basic courses in ____.
    A. English
    B. math
    C. science
    D. All of the above.

7. Which of these statements is true of apprentices?
   A. They do not receive pay until the apprenticeship is completed.
   B. They spend part of their time in classes and part working in the trade.
   C. They spend all of their time working with a master carpenter.
   D. None of the above.

_____Seven_____

8. Historically, the carpenter's apprentice usually lived in the master carpenter's household and received training over a time span as long as ____ years.
   A. four
   B. five
   C. six
   D. seven

_____Journeyman_____

9. Historically, when training was complete and the master felt that the apprentice had attained a high level of skill, the apprentice was granted the status of ____ and could then work for wages.

_____management_____

10. Today, apprenticeship training programs are carefully organized and supervised. Local committees representing labor and ____ provide direct control.

_____17_____

11. Applicants for apprentice training programs in carpentry must be at least ____ years old and must satisfy the local committee that they have the ability to master the trade.
    A. 16
    B. 17
    C. 18
    D. 19

_____four_____

12. The term of apprenticeship for carpentry is normally ____ years, but may be reduced for applicants who have completed advanced courses in vocational-technical schools.
    A. two
    B. three
    C. four
    D. five

_____144_____

13. In addition to instruction and skills learned on the job, an apprentice must attend classes in subjects related to carpentry. Classes are usually held in the evening and must total at least ____ hours per year.
    A. 72
    B. 108
    C. 120
    D. 144

_____50%_____

14. The wage scale for an apprentice is determined by the local apprenticeship committee and usually starts at about ____% of the journeyman's scale during the preceding year.
    A. 40
    B. 45
    C. 50
    D. 55

Journeyman   15. When the training period is complete and apprentices have passed a final exam, they are issued a certificate stating that they are ____ carpenters.

Entrepreneurs   16. People who start and operate their own businesses are called ____.

_____   17. Which skill or behavior do most contractors consider most important when hiring a carpenter?
A. Having respect for customers and owners.
B. Using leveling instruments.
C. Operating a screw gun.
D. Installing door hardware.

Read/study new books Code changes, safety   18. List three things that make it important to keep learning after you become a journeyman.

19. What organization exists to hold state and national competition for students of the building trades?

Skills USA

_____   20. Which organization concentrates on building homes?
A. Associated General Contractors.
B. National Association of Home Builders.
C. United Brotherhood of Carpenters and Joiners of America.
D. Home Builders International.

CHAPTER **2**

Safety

Carefully study the chapter and then answer the following questions.

_____   1. OSHA requires employers to provide a workplace that is free from recognized ____.

_____   2. According to OSHA, a competent person is one who:
   A. has read the manuals and passed a state test.
   B. is capable of identifying hazards and is authorized to correct them.
   C. has enough experience to recognize when something is unsafe.
   D. is an OSHA employee responsible for correcting hazardous conditions.
   E. has taken appropriate classes and passed the OSHA test.

_____   3. Under OSHA, a worker is responsible for:
   A. being familiar with the OSHA poster that is posted in every workplace.
   B. following his or her employer's safety rules.
   C. using all safety gear that is required.
   D. reporting all safety hazards to his or her supervisor or the safety committee.
   E. All of the above.

_____   4. Remove ____ from scraps of lumber before discarding them.

_____   5. ____ fitting clothing can pull hands or other body parts into cutting tools.

_____   6. Shoes with ____ soles should never be used in carpentry; they have poor traction on smooth surfaces.

_____   7. Approved safety glasses should be stamped with ANSI number ____.
   A. Z81.7
   B. Z78.1
   C. Z87.1
   D. Z88.7

_____ 8. Hard hats must be able to resist breaking from the force of a(n) _____ dropped from a height of 5′.
A. 8 lb. ball
B. 80 lb. ball
C. 80 oz. ball
D. 1/8″ diameter steel ball

_____ 9. _True or False?_ A particulate mask must be worn whenever you are removing paint.

_____ 10. _True or False?_ A respirator must be worn whenever you are removing asbestos.

_____ 11. Fall protection is required if you work within _____ feet of the edge of a surface that is more than six feet from the ground or a stable surface.

_____ 12. Workers on a roof with a pitch of _____ or steeper must wear safety harnesses.

_____ 13. Scaffolds should be inspected every _____ for safe conditions.
A. week
B. day
C. time erected
D. month

_____ 14. A scaffold with a load rating of 400 pounds must be capable of supporting a load of _____.
A. 400 pounds
B. 400 pounds plus the weight of a worker
C. 800 pounds
D. 1,600 pounds

_____ 15. Which is _not_ a safety rule for hand tools?
A. Hold edge tools with both hands and the cutting action away from your body.
B. Keep edge and pointed tools pointed downward.
C. Carry edge and pointed tools in your pocket with the sharp edge pointed downward.
D. Store tools in a proper box, case, or cabinet when not in use.

16. Which conductor on a power tool cord normally carries the current back to its source?

_____

_____ 17. Which of these devices is most effective at protecting against electric shock?
A. GFCI.
B. Disconnect switch.
C. Circuit breaker.
D. Fuse.

_____ 18. _True or False?_ Excavations can be safely entered if the sides are sloped to their angle of repose.

Name _____

_____ 19. What phone number can be called to find the locations of underground utilities?

_____ 20. What does SDS stand for?
A. Structural Design System.
B. Safety Data Sheet.
C. System Disconnect Switch.
D. None of the above.

_____ 21. How should CCA-treated wood be disposed of?
A. Burn with other combustible waste.
B. Run it through a wood chipper and use it as compost.
C. Dispose of it with yard waste.
D. None of the above.

_____ 22. Carpenters should know that special methods must be used to control fires caused by electrical wiring or equipment. This type of fire is referred to as a Class ____.
A. A
B. B
C. C
D. None of the above.

_____ 23. What type of fire is burning wood and paper?
A. Class A.
B. Class B.
C. Class C.
D. None of the above.

24. Look at the illustration below and list the safety measures the person is taking.

A. _____

B. _____

C. _____

D. _____

Name _Matthew Horner_   Date _08/27/18_   Score _____

CHAPTER **3**

Building Materials

Carefully study the chapter and then answer the following questions.

Lignin   1. The basic structure of wood consists of narrow tubes or cells held together with a natural substance called ____.

2. Identify the basic parts of the tree trunk/log shown in the illustration below.

    A. _Heartwood_
    B. _Sapwood_
    C. _Wood Rays_
    D. _Cambium_
    E. _Bark_
    F. _Annular Rings_

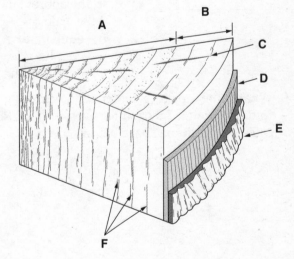

Summer   3. Wood cells are formed in the cambium layer. This growth occurs in the spring and ____ seasons.

Conifers   4. Softwood comes from the evergreen, or needle-bearing, trees. These trees are called ____ because many of them bear cones.

_J, B, A, C_   5. Listed below are several commonly used hardwoods and softwoods. Which four are hardwoods?
    (A.) White ash
    (B.) Basswood
    (C.) Birch
    D. Cypress
    E. Douglas fir
    F. Hemlock
    G. Pine
    H. Redwood
    I. Spruce
    (J.) Willow

_____B_____   6. Open-grain wood refers to wood that ____.
  A. has a wide grain pattern usually used in damp locations
  B. is made up of large cells with openings that require extra steps in finishing
  C. absorbs moisture more quickly than other species and should not be used in wet locations
  D. has heartwood at the surface, creating attractive grain patterns for fine woodwork
  E. has sapwood at the surface, making it less attractive for use in fine woodwork
  F. is grown in open areas where it gets plenty of sunlight and grows quickly

edge-grained   7. When softwood lumber is cut so the annular rings form an angle of more than 45° with the surface of the board, the lumber is called ____.
  A. quarter-sawed
  B. flat-grained
  C. plain-sawed
  D. edge-grained

equilibrium   8. In the abbreviation E.M.C., the E stands for ____.
  A. elasticity
  B. equilateral
  C. equalized
  D. equilibrium

9. What is the moisture content of a board if a test sample that originally weighed 8.5 oz. was found to weigh 7.5 oz. after oven drying? Round your answer to the nearest percent. Show your calculations in the space below.

$$M.C. = \frac{IW - OW}{OW} = \frac{8.5 oz - 7.5 oz}{7.5 oz} = \frac{1}{7.5 oz} = 13.3 =$$

13% = M.C.

_30%_   10. For most woods, the fiber saturation point is about ____%.
  A. 20
  B. 24
  C. 30
  D. 36

Name _____

_____8%_____ 11. The average moisture content of interior woodwork (after it reaches equilibrium) installed in homes located in the Midwestern United States is about ____%.
A. 6
B. 8
C. 11
D. 15–19

_____Oven_____ 12. Moisture content of wood can be determined by ____ drying or using an electronic moisture meter.

13. Embedded branches or limbs of the tree cause knots in lumber. Identify the common types shown below.

A. Sound intergrown through two faces
B. Sound star checked, intergrown through 2 wide faces
C. Sound encased through two wide faces.
D. Round Knothole though two wide faces.

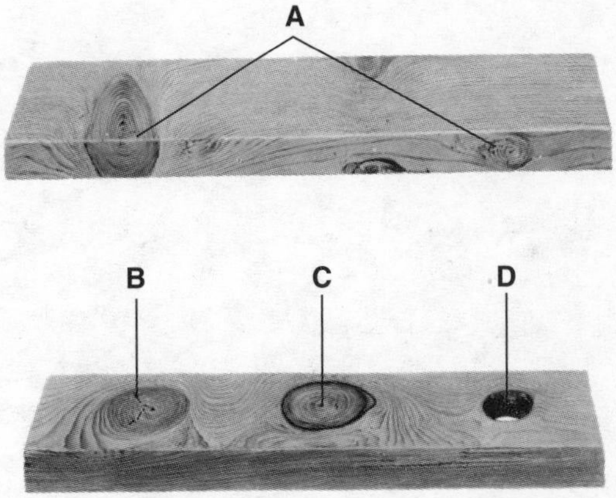

_____C_____ 14. Wood defects consisting of separation across the annular rings are called ____.
A. checks
B. splits
C. Both A and B.
D. None of the above.

_____Shakes_____ 15. When the separation occurs between the annular rings, the defects are called ____.

16. Any variation in a board from a true or plane surface is generally referred to as warp. Identify the specific kinds of warp illustrated below.

A. _____Crook_____

B. _____twist_____

C. _____bow_____

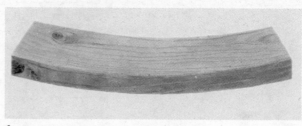

A

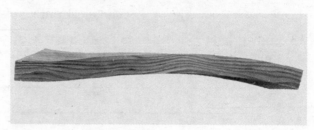

B

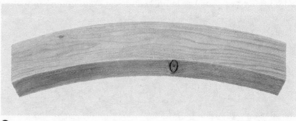

C

_____dimensions_____ 17. The three basic classifications of softwood lumber include boards, ____, and timbers.

_____premium_____ 18. A special or ____ grade of hardwood is known as "architectural" or "sequence-matched."

_____FAS_____ 19. The best grade of hardwood lumber normally available is identified by the letters ____.

Name _____

20. Rough lumber is surfaced and otherwise machined before it is used. Identify the width classifications in the view below.

A. _Nominal width_____

B. _dressed width_____

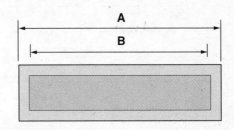

__A_____ 21. The actual width of a standard, dried 2 × 8 piece of S4S framing lumber is ____".

Ⓐ 7 1/4
B. 7 3/8
C. 7 1/2
D. 7 5/8

22. A stack of 2 × 8s contains 21 pieces 16′ long, 12 pieces 14′ long, and 10 pieces 12′ long. What is the total bd. ft. of the stack? Show your calculations in the space below.

$$bd.ft = \frac{No. Pcs. \times T \times W \times L}{12}$$ =   43   42

$$\frac{21 \times 2 \times 8 \times 16}{12} = 448 \qquad \frac{12 \times 2 \times 8 \times 14}{12} = 224$$

$$\frac{10 \times 2 \times 8 \times 12}{12} = 160 \qquad 448 + 224 + 160 = \boxed{832\ bd.ft}$$

_Southern Pine_____ 23. Species (kinds) of softwood used in manufacture of plywood are classified in five groups. Group 1 represents the highest level of stiffness. Which one of the following is included in this group?
A. Redwood.
Ⓑ Southern pine.
C. Spruce.
D. Western hemlock.

_Span Rating_____ 24. ____ is a plywood strength rating and indicates the greatest recommended center-to-center distance in inches between supports when the long dimension of the panel is at right angles to the supports.

25. Note the illustration of an APA-The Engineered Wood Association trademark and indicate the meaning of each item designated with a letter. Place your answers in the appropriate spaces provided at the right.

A. _____Panel grade_____

B. _____Span Rating_____

C. _____Bond classification_____

D. _____Mill number_____

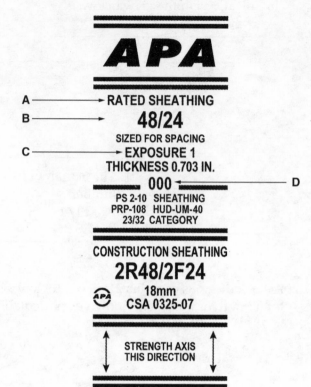

_____C_____
26. Which of the following is the best description of OSB?
    A. A designation for lumber that is larger than standard dimensions, meaning oversize board.
    B. A designation for panels that must be supported on structural bearing.
    C. The name of a panel material that is made up of strands or flakes of wood oriented in the same direction in layers with the layers being held together by synthetic resin.
    D. An engineered product for use on ceilings called overhead structural boards.

_____40_____
27. Particleboard is made from flakes, chips, and shavings bonded together with adhesives. Average panels weigh _____ lb. per cu. ft.
    A. 35
    B. 40
    C. 45
    D. 50

_____5/8"_____
28. Hardboard is made from refined wood fibers pressed together to form a hard, dense material. Hardboard is available in thicknesses up to _____".
    A. 1/4
    B. 5/16
    C. 1/2
    D. 5/8

Name _____

False

29. *True or False?* OSB cannot be used as an exterior siding.

Oriental Strand board

30. In the building material OSB, what do the letters OSB stand for?
    A. Optional Sheathing Board
    B. Oriented Sheathing Board
    C. Oriented Strand Board
    D. Outside Structural Board

offset

31. The prefabricated structural unit indicated by the arrow is known as a(n) ____.

Tongue-and-groove lumber

32. Which of the following is *not* an engineered lumber product?
    A. Glue-laminated beams
    B. Laminated veneer lumber
    C. Open-web joists
    D. Tongue-and-groove lumber

33. What are the advantages of engineered lumber over conventional lumber products?
    A. Engineered lumber can be stronger.
    B. Engineered lumber can be straighter.
    C. Engineered lumber can be more dimensionally stable.
    D. All of the above are possible.

34. What materials are usually used to manufacture I-joists?
    A. OSB for the web and solid lumber for the flanges.
    B. Plywood for the web and the flanges.
    C. Solid lumber for the web and OSB or plywood for the flanges.
    D. Solid lumber for all parts.

35. Nails are the most commonly used metal fastener. Identify the basic types shown below.
    A. _Common_
    B. _Box_
    C. _Casing_
    D. _Finish_

_3 1/2"_

36. What is the length of a 16d nail?
    A. 3"
    B. 3 1/4"
    C. 3 1/2"
    D. 3 3/4"

Spiral threads 37. ____ are done in the manufacture of nails to increase their holding power.

38. Wood screws have greater holding power than nails. Identify the three standard types of slotted screws shown below.

   A. *Round head, straight slot*
   B. _____
   C. *Flat head, straight slot*

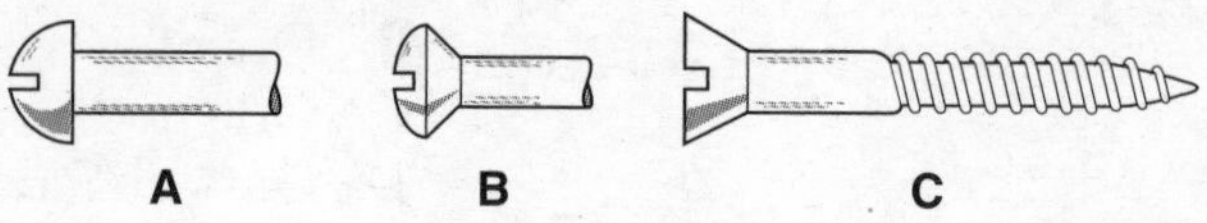

A   **B**   **C**

_____ 39. Polyvinyl resin emulsion adhesive can be used to assemble a cabinet drawer. Which one of the following statements concerning the characteristics of this kind of glue is *not* correct?
   A. The resin material used in the glue is thermoplastic.
   B. The glue has a high resistance to moisture.
   C. The glue is available in ready-to-use form.
   D. The glue sets up rapidly and does not stain the wood.

More   40. Thermoset adhesives are (*more, less*) ____ water-resistant than polyvinyl adhesives.

Contact Cement   41. An adhesive called ____ is an excellent adhesive for applying plastic laminates.
   A. contact cement
   B. polyvinyl glue
   C. casein glue
   D. urea cement

Name _____ Date_____ Score _____

CHAPTER **4**

Hand Tools

Carefully study the chapter and then answer the following questions.

_____ 1. Two sizes of tape measures that many carpenters carry are ____.
    A. 25-foot and 100-foot
    B. 6-foot and 40-foot
    C. 50-foot and 200-foot
    D. 20-foot and 75-foot
    E. 3-foot and 40-foot

_____ 2. In the illustration below, the measuring tool being used to mark a rafter is called a(n) ____.

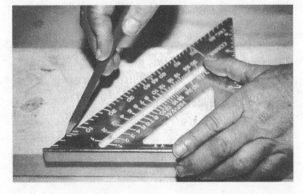

_____ 3. Which is the best tool to use to mark a straight line 20 feet long on a floor?
    A. Long level.
    B. Straight piece of lumber.
    C. Chalk line.
    D. 25-foot tape measure.
    E. Transit.

_____ 4. On a rafter square, the ____ is at a right angle to the blade.

_____ 5. Which of the following operations *cannot* be done with a speed square?
    A. Mark square cut-off lines.
    B. Mark any angle from 0° to 90°.
    C. Layout rafters.
    D. Guide a circular saw in making square cuts.
    E. All of the above can be done.

_____ 6. The ____ has a movable blade, making it possible to transfer an angle from one place to another.

_____ 7. What is the name of the level that can be hung on a tight string to check for levelness over long distances?
    A. Line.
    B. Stretch-out.
    C. Mini.
    D. String.
    E. None of the above.

8. In the illustration below are five basic types of handsaws used by the carpenter. Correctly identify each saw.

    A. _____

    B. _____

    C. _____

    D. _____

    E. _____

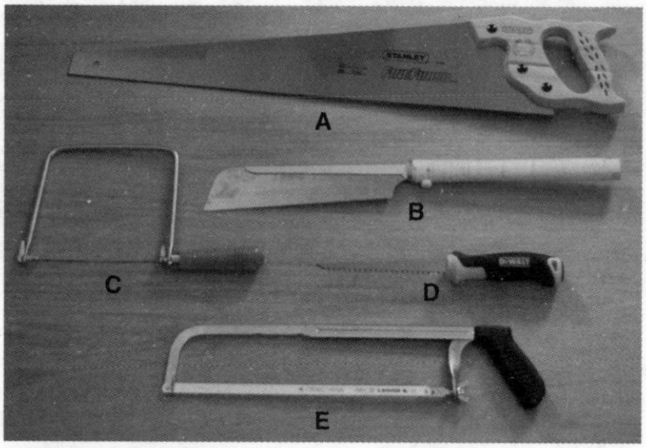

_____ 9. The cut made by a saw is called a(n) ____.

10. The tooth size of a handsaw is specified by listing the points per inch. How many teeth per inch are found on a crosscut saw?

_____

_____ 11. A jack plane is commonly selected for general purpose work. The bed of this plane is ____″ long.
    A. 8 to 9
    B. 10
    C. 12
    D. 14

_____ 12. A(n) ____ plane has a blade mounted at a low angle with the bevel of the cutter turned upward.

_____ 13. A(n) ____ or soft-faced hammer should be used to strike a chisel when making deep cuts.

14. Identify the parts of the standard claw hammer shown below.

    A. _____

    B. _____

    C. _____

    D. _____

    E. _____

    F. _____

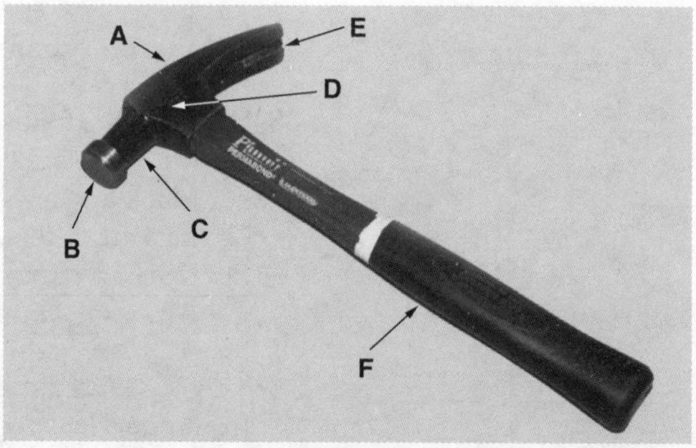

Name _____

15. What type of striking tool is pictured below?

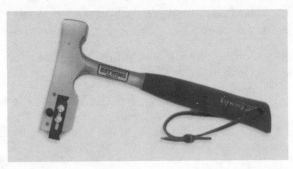

_____

_____ 16. Nail sets are about 4″ long. Size is determined by the diameter of the tip. Sizes vary in _____″ increments.
    A. 1/64
    B. 1/32
    C. 1/16
    D. 1/8

_____ 17. Standard screwdriver size is determined by the length of the blade measured from the base of the _____ to the tip.
    A. ferrule
    B. handle
    C. head
    D. sleeve

18. Identify the four types of screwdrivers shown below.

    A. _____

    B. _____

    C. _____

    D. _____

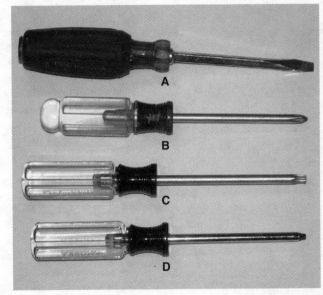

_____ 19. How are the sizes of hand screws designated?
                            A. Width of the jaw opening.
                            B. Length of the jaws.
                            C. Maximum force that can be applied.
                            D. C-clamp.

20. The drawing below shows an edge view of a standard plane iron. Give the recommended angles for honing and grinding.

A. _____

B. _____

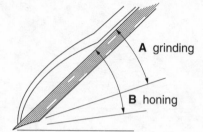

_____ 21. ____ is a saw sharpening operation in which the points of the teeth are filed to equal height.
                            A. Leveling
                            B. Striking
                            C. Setting
                            D. Jointing

22. Look at the illustration below and describe the tool maintenance being done.

Name _____   Date _____   Score _____

CHAPTER **5**

Power Tools

Carefully study the chapter and then answer the following questions.

_____ 1. The two general types of power tools are portable and ____.

_____ 2. Electrical shock is one of the potential hazards in the operation of portable tools. In addition to using approved receptacles, plugs, and cords, always be sure that the tool is properly ____.

_____ 3. A ____ detects very slight differences in the current flowing in the hot conductor and the neutral conductor.
   A. circuit breaker
   B. fuse
   C. circuit monitor
   D. ground fault circuit interrupter

_____ 4. Under normal conditions, electric current flows through:
   A. the hot conductor.
   B. the neutral conductor.
   C. the grounding conductor.
   D. both the hot and neutral conductors.
   E. all three conductors.

5. Identify the specified parts of the circular saw shown below.

   A. _____
   B. _____
   C. _____
   D. _____
   E. _____
   F. _____

_____ 6. The depth of cut of the portable circular saw is adjusted by raising or lowering the ____.

_____ 7. The depth of cut of the portable circular saw should be adjusted so the blade cuts through the stock and projects about ____".
A. 1/8
B. 1/4
C. 3/8
D. 1/2

_____ 8. The type of saw shown below is a ____ saw.
A. worm-drive
B. speed
C. push
D. timber

_____ 9. A saber saw blade cuts on the ____ (*upward, downward*) stroke.

_____ 10. The length of stroke of a saber saw blade is about ____".
A. 1/4
B. 1/2
C. 3/4
D. 1

_____ 11. The size of a portable electric drill is determined by the ____.
A. horsepower of the motor
B. highest rpm capability
C. capacity of the chuck
D. length of the drill body

_____ 12. Which of the following is *not* a valid safety rule for the portable electric drill?
A. Stock to be drilled must be held in a stationary position.
B. Place the base of the drill firmly on the stock before starting the motor.
C. When drilling deep holes with a twist drill, withdraw the drill several times to clear cuttings.
D. Always remove the drill bit from the chuck as soon as you have completed your work.

Name _____

13. Identify the types of drill bits shown below.

  A. _____

  B. _____

  C. _____

  D. _____

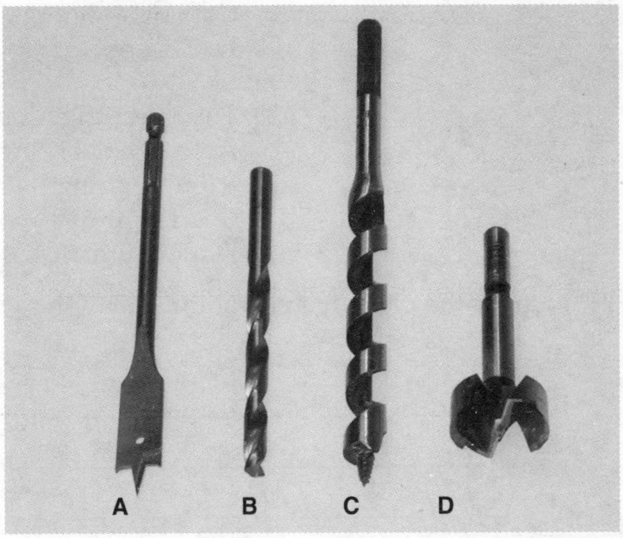

14. The power plane motor drives a spiral cutter at speeds of about
   ____ rpm.
   A. 5000
   B. 10,000
   C. 15,000
   D. 20,000

15. The depth of the cut of a power plane is adjusted by raising or
   lowering the ____.
   A. front shoe
   B. rear base
   C. cutter head
   D. motor assembly

16. Which of the following *cannot* be cut with a router?
   A. Dadoes.
   B. Grooves.
   C. Contours and curves.
   D. Dovetail joints.
   E. All of the above can be done with a router.

17. When viewed from above, a router motor revolves in a ____
   (*clockwise, counterclockwise*) direction.

18. The size of a portable belt sander is determined by the ____.
   A. overall width
   B. belt width and length
   C. overall length
   D. motor hp

_____ 19. Which of the following is a common type of pneumatic nailer?
    A. Battery-operated.
    B. Strip-fed.
    C. Single shot.
    D. Gas-operated.

_____ 20. The safest trigger setting for a power nailer is ____.
    A. sequential firing
    B. bump firing
    C. safety fire
    D. slow-fire

21. Identify the parts of a power miter saw shown below.

A. _____

B. _____

C. _____

D. _____

E. _____

22. What should be done before changing blades on a power miter saw or chop saw?

_____

_____ 23. To adjust the depth of cut of a standard radial arm saw, the ____ is raised or lowered.
    A. table base
    B. motor mounting
    C. overhead arm
    D. vertical column

24. The drawing below shows a standard setup for crosscutting with the radial arm saw. Identify the parts, thrust direction, and saw feed.

A. _____

B. _____

C. _____

D. _____

E. _____

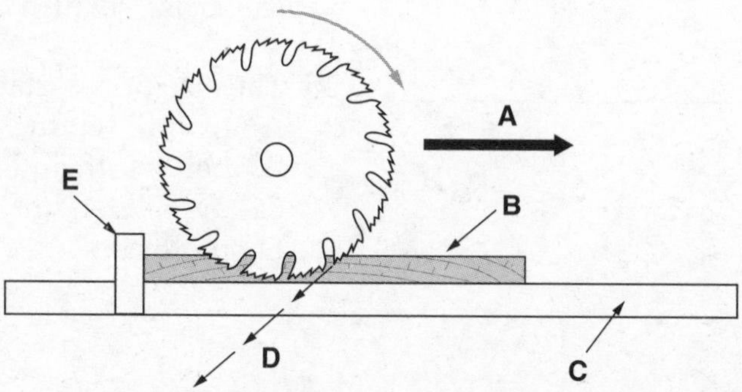

Name _____

_____ 25. Safety rules listed in the text prescribe a ____" margin of safety when operating the radial arm saw.
A. 2
B. 4
C. 6
D. 8

_____ 26. When performing ripping operations on a radial arm saw, always feed the stock into the blade so that the bottom teeth are turning ____ (*toward*, *away from*) you.

_____ 27. When ripping stock on a table saw, the stock must have a(n) ____ face to rest on the table and a straightedge to run against the fence.

_____ 28. Which of the following is *not* a valid safety rule for operating a standard table saw?
A. Set the blade so it extends 1/4" above the stock to be cut.
B. When ripping stock freehand, do not use the fence.
C. Always use push sticks when ripping short, narrow pieces.
D. Maintain a 4" margin of safety even when the guard is in position.

29. Identify the specified parts of the standard jointer illustrated below.

A. _____

B. _____

C. _____

D. _____

E. _____

F. _____

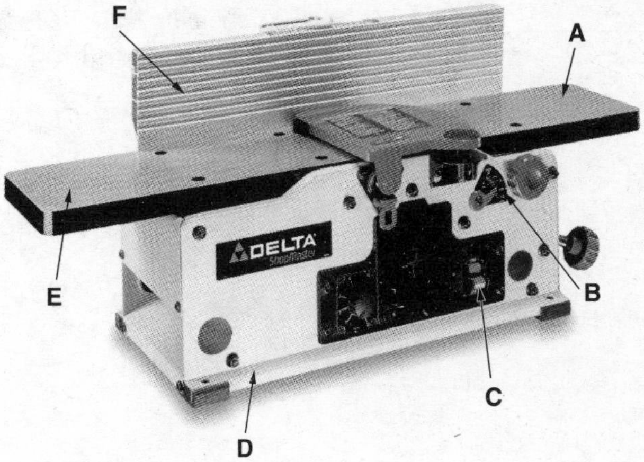

_____ 30. When jointing the edge of a straight piece of stock, the depth of cut will gradually decrease (forming a taper) if the ____.
A. cutter head is adjusted too high
B. outfeed table is slightly higher than the cutter head
C. infeed table is too low
D. outfeed table is lower than the infeed table

_____ 31. The tool illustrated below, designed to attach drywall with screws, is known as a(n) ____.

_____ 32. A powder-actuated driver uses ____ to drive fasteners into concrete and steel.
A. compressed air
B. exploding powder
C. a gas cartridge
D. an electrical discharge

_____ 33. *True or False?* Most power tools require frequent lubrication with grease or oil to keep them in good condition.

CHAPTER **6**

Scaffolds, Ladders, and Rigging

Carefully study the chapter and then answer the following questions.

_____ 1. The major causes of scaffold accidents are ____.
   A. slipping
   B. support giving way
   C. being struck by a falling object
   D. All of the above.

_____ 2. The height of the scaffolding platform is important, since it must permit the work to be performed with speed and accuracy without causing unnecessary stooping or ____ by the worker.

_____ 3. Many builders use steel or ____ scaffolding.

4. Typical designs for single-pole and double-pole wooden scaffolding are shown in the drawings below. Identify the specified parts.

A. _____
B. _____
C. _____
D. _____
E. _____
F. _____
G. _____

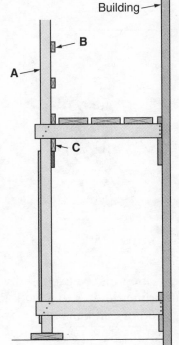

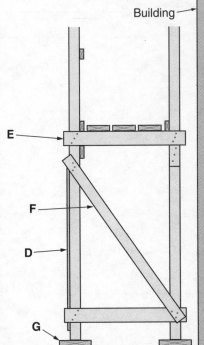

_____ 5. Horizontal members used to connect scaffold uprights are called ____.

_____ 6. When erecting either single-pole or double-pole scaffolding, ledger sections consist of 2 x 6s about ____′ apart.
   A. 4
   B. 6
   C. 10
   D. 12

_____ 7. Sectional steel or aluminum scaffolds are commonly used. Sections can be rapidly assembled from truss frames and ____.

_____ 8. All planked areas of scaffolds are required to have guardrails. The tops of guardrails must be between ____ inches above the planks.
   A. 6–12
   B. 15–20
   C. 18–24
   D. 39–45

_____ 9. All planked areas of scaffolds must have a toeboard to prevent tools and materials from slipping off the platform. The toeboard must be within ____″ of the top of the planks.
   A. 1/2
   B. 1
   C. 2
   D. 3

10. The drawings below show three types of metal devices commonly used in connection with wood planks to form low-level scaffolds. Identify each item shown.

   A. _____

   B. _____

   C. _____

A   B   C

_____ 11. Great care must be exercised when using nails to attach metal scaffolding devices to a wall. Recommendations suggest the use of ____ 16d or 20d nails driven into sound framing lumber.
   A. two
   B. three
   C. four
   D. five

Name _____

_____ 12. Metal units commonly used to support low platforms for interior work are called _____. They are attached to a wooden ledger and can be adjusted to several different heights.

13. Provide the correct names of the ladders shown below.

    A. _____

    B. _____

    C. _____

    D. _____

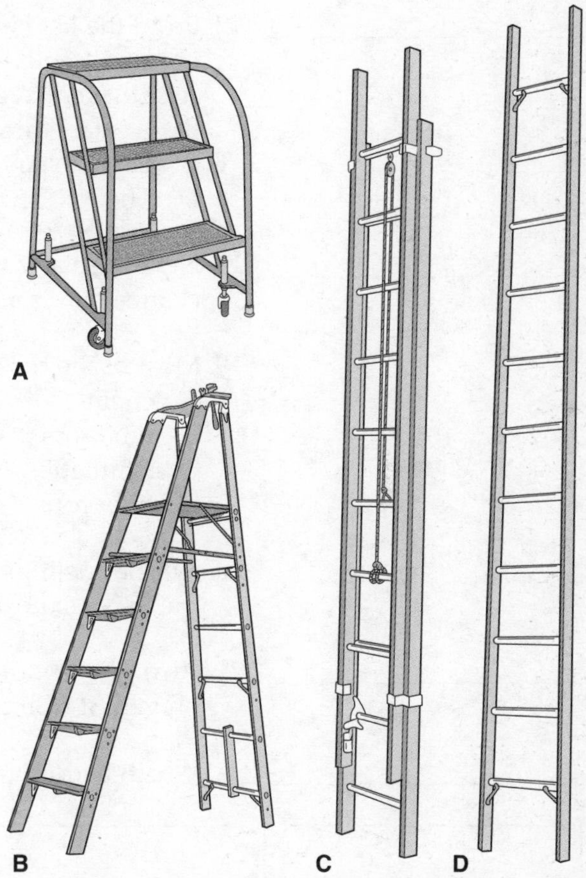

A

B **C** **D**

_____ 14. When placing a ladder against a wall, the horizontal distance from the lower end to the wall should be at least _____% of the total length of the ladder.
    A. 15
    B. 20
    C. 25
    D. 30

_____ 15. When a ladder is used to gain access to a roof, it should be long enough to extend above the roof a distance of at least _____'.
    A. 2 1/2
    B. 3
    C. 3 1/2
    D. 4

_____ 16. When a ladder is used on surfaces that may permit the bottom end to slip, it is best to _____.
    A. nail the rails to anti-skid boards
    B. place the bottom rails of the ladder in a special frame
    C. equip the bottom end of the rails with safety shoes
    D. use a rope to hold the bottom end in place

_____ 17. From the list below, select the *incorrect* statement concerning the safe use of a stepladder.
    A. Do not leave tools on the surface of the top step.
    B. At least three legs should rest on solid support.
    C. Do not stand on the two top steps.
    D. The two sections should be fully opened and locked.

_____ 18. Never use any type of metal ladder where there is the slightest chance that it might make contact with _____.

_____ 19. Most of the hoisting used by carpenters is with _____ slings.
    A. chain
    B. wire rope
    C. synthetic
    D. fiber rope

_____ 20. Synthetic slings are *not* damaged by water, but they are damaged by excessive exposure to heat and _____.

_____ 21. A(n) _____ should be attached to a load to be hoisted by a crane to prevent it from twisting and swinging.

_____ 22. Which hand signal in the illustrations below indicates emergency stop?

CHAPTER 7

Plans, Specifications, and Codes

Carefully study the chapter and then answer the following questions.

1. From the roof-framing plan shown below, provide this information:

   A. _____ Size of rafters (include all sizes).

   B. _____ Where furring material is used.

   C. _____ Spacing of rafters.

   D. _____ Dimensions of ridge boards.

   E. _____ Where 2 × 4s are used.

   F. _____ Support material used under furred rafters.

   G. _____ Spacing of collar ties.

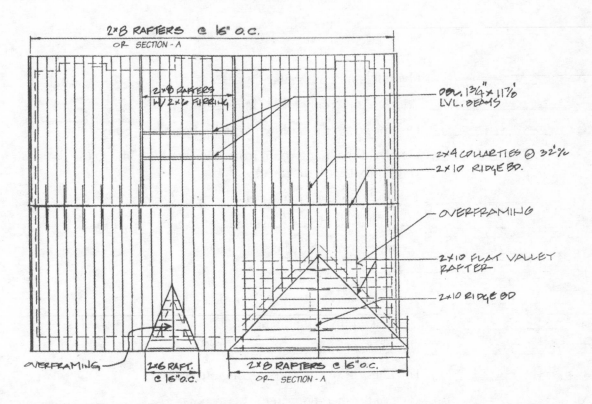

ROOF FRAMING PLAN
1/8" = 1'-0"

2. Study the floor plan and provide the following information:

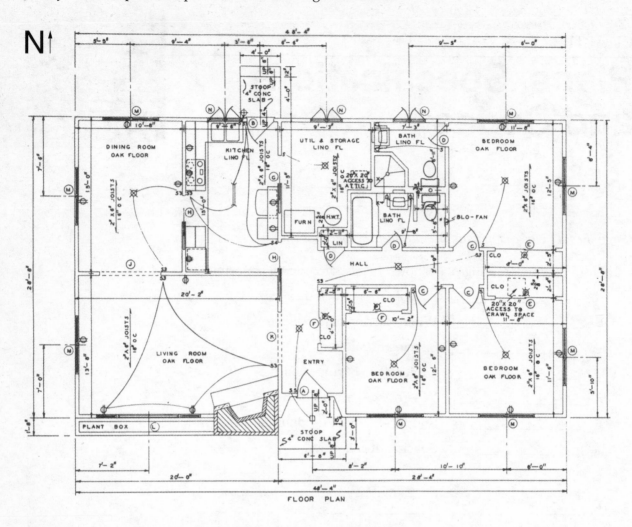

FLOOR PLAN

A. _____ Overall length of the west wall.

B. _____ Width of the front stoop concrete slab.

C. _____ Distance from the bath window to the nearest corner.

D. _____ Length of the living room from the sheathing line to the partition center (east-west distance).

E. _____ The width of the kitchen between wall centers.

F. _____ Material used for the finished floor in the utility room.

G. _____ Cross-sectional size of ceiling joists.

H. _____ Spacing specified for ceiling joists.

I. _____ Size of the access door to the attic.

J. _____ Outside width of the plant box.

K. _____ Number of bathtubs.

L. _____ Thickness of the rear stoop concrete slab.

M. _____ Type of finished floor in the bedrooms.

Name _____

3. The architect often uses symbols to indicate certain materials. Identify the basic materials represented by the symbols shown below.

Material	Plan	Elevation	Section
A	Floor areas left blank	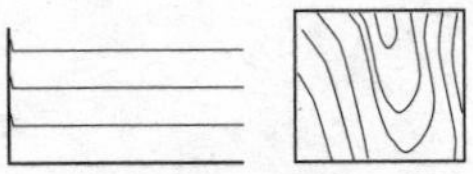 Siding Panel	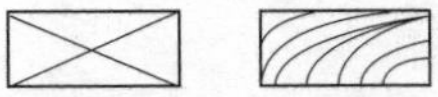 Framing Finish
B	 Face 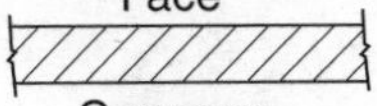 Common	 Face or common	Same as plan view
C	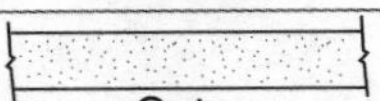 Cut 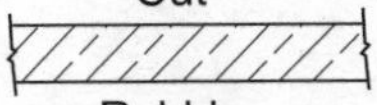 Rubble	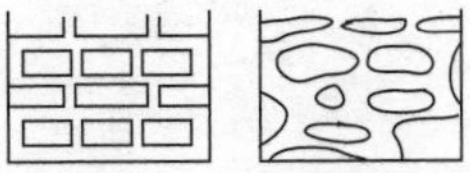 Cut Rubble	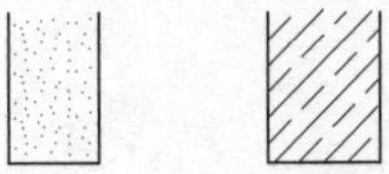 Cut Rubble
D	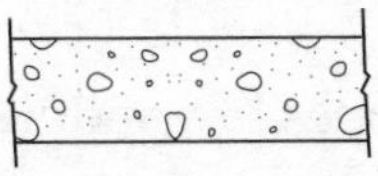	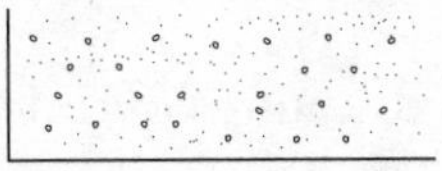	Same as plan view
E	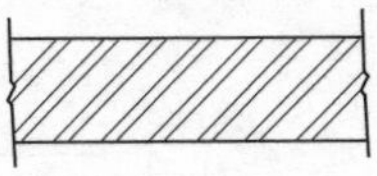	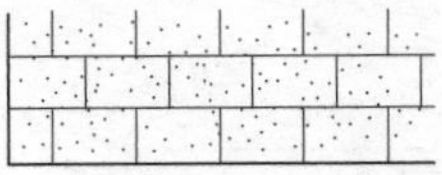	Same as plan view
F	None	None	
G		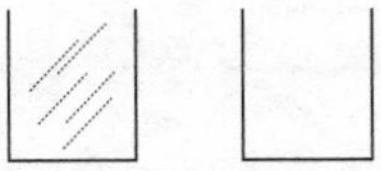	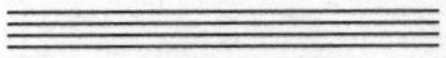 Large scale Small scale

A. _____   E. _____

B. _____   F. _____

C. _____   G. _____

D. _____

4. For most measurements over 2' or 3' long, carpenters prefer to work with feet and inches. Find the total of the following measurements: 7'-4", 3'-6", 10'-10", 12'-5", 9'-10", and then subtract 69". Show your calculations in the space below.

5. Below are eight plan view drawings of different types of windows, doors, and other openings. Identify each one.

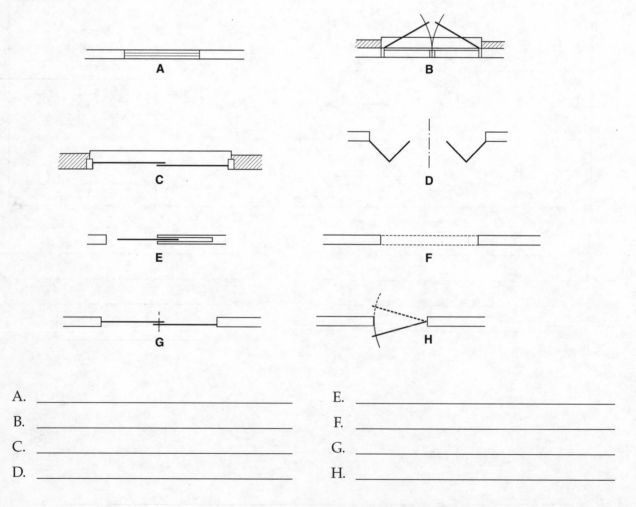

A. _____   E. _____

B. _____   F. _____

C. _____   G. _____

D. _____   H. _____

Name _____

6. Simplified drawings and symbols are used to show plumbing fixtures, appliances, and mechanical equipment. Identify the following symbols.

A. _____

B. _____

C. _____

D. _____

E. _____

F. _____

G. _____

H. _____

A

B

C

D

E

F

G

H

_____ 7. In written specifications, under the heading Painting and Finishing, which of the following would *not* be appropriate?
A. Surfaces to be included (interior and exterior).
B. Specifications of materials to be used.
C. Application method and number of coats.
D. Number of painters to be used on the job.
E. Guarantee of quality and performance.
F. Completion date.

_____ 8. Which of the following metric units is closest to a US customary yard in length?
A. Kilometer.
B. Meter.
C. Centimeter.
D. Millimeter.

_____ 9. Commercial standards are developed by the Commodity Standards Division of the U.S. Department of ____.

_____   10. To secure a building permit, the contractor or _____ must file an application along with drawings and specifications.

_____   11. When work gets underway at the building site, a special card furnished by the local building department must be posted. As each stage of construction is completed, the card is signed by the _____.
A. contractor
B. owner
C. inspector
D. carpenter

CHAPTER 8

Building Layout

Carefully study the chapter and then answer the following questions.

_____ 1. A(n) ____ indicates the exact location of the structure to be built, as well as the distances to the property lines.

2. Two standard types of measuring tapes are shown in the illustration below. What are the readings designated at A and B?

A. _____

B. _____

_____ 3. The operation of leveling instruments is based on the fact that a line of sight is always a(n) ____ line.

_____ 4. Which of the following operations *cannot* be performed with a standard builder's level?
   A. Checking plumb lines.
   B. Laying out horizontal angles.
   C. Measuring horizontal angles.
   D. Laying out level lines.

5. Identify the parts of the standard builder's level shown below.

A. _____

B. _____

C. _____

D. _____

E. _____

_____   6. When in use, the level-transit must be securely mounted on a ____.
  A. transit stand
  B. leveling rod
  C. tripod
  D. rod base

7. The view below shows how the leveling screws are adjusted on a standard instrument. With the screws being turned as indicated by the arrows, will the bubble move in the direction shown in A or the direction shown in B?

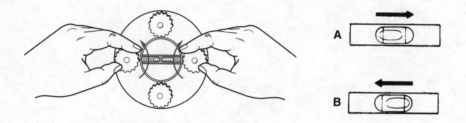

8. The view below shows a builder's level positioned in two locations on a steep slope. Given the rod readings presented, what is the total difference in elevation between A and C? Show your calculations in the space below.

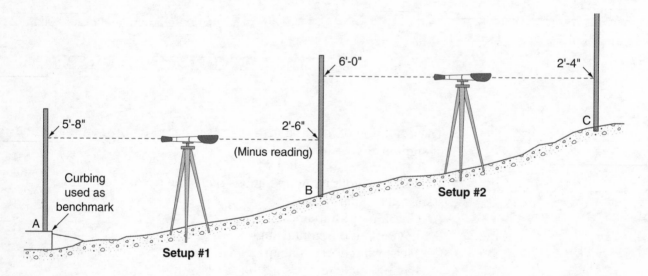

Name _____

9. The vernier scale is used to lay out or measure angles that include fractions of a degree. What is the exact angle shown below?

_____

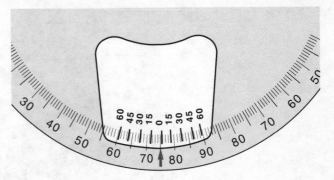

10. How many minutes are in a complete circle? How many seconds are in an angle of 15°? Use the space below for calculations.

Minutes _____

Seconds _____

_____   11. A straight line that does not dip, sag, or curve is called a(n) ____.

_____   12. The ____ (*builder's level, transit*) can be pivoted vertically in each direction.

_____   13. When sighting through the telescope of a leveling instrument, it is recommended that ____.
A. one eye be kept closed
B. both eyes be kept open

_____   14. A(n) ____ is marked off with numbered graduations and should be used with a transit when sighting over a long distance (over 100′).

_____   15. An officially established reference point that can be used for all elevations on the building is called a(n) ____.

16. Look at the illustration below. Explain how attaching a laser level receiver to an excavating machine can accelerate excavating.

_____

_____

_____

_____

_____

Name _____ Date _____ Score _____

CHAPTER **9**

Footings and Foundations

Carefully study the chapter and then answer the following questions.

_____ 1. To protect the owner and builder, the assistance of a registered engineer or licensed surveyor should be secured when laying out or checking ____.

2. When a laser level, builder's level, or transit is *not* available, a 90° right angle can be established by measured distances. Referring to the drawing below, provide the measurements commonly used in this method.

A. _____

B. _____

C. _____

D. _____

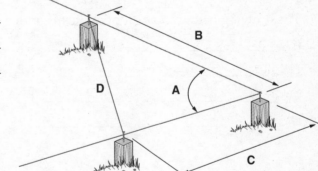

_____ 3. After building lines are established, it is good practice to check the length of ____ that exist in squares and rectangles to be sure that corners are square.

_____ 4. A batter board assembly consists of stakes and one or more horizontal members. Each horizontal member is called a ____.
A. bracket
B. brace
C. leveling strip
D. ledger

_____ 5. For regular basement foundations in residential construction, the excavation should extend beyond the building lines by at least ____′.
A. 2
B. 3
C. 4
D. 5

_____ 6. Before beginning any excavation, call _____ for help in determining where any underground utilities are located.
  A. 911
  B. 811
  C. 411
  D. *555

7. Provide the recommended footing dimensions in the drawing below. Consider the foundation wall to be 10" thick.

  A. _____

  B. _____

  C. _____

  D. _____

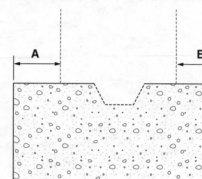

_____ 8. For a two-story residential structure, it is recommended that chimney footings have a minimum thickness of _____".

9. The drawing below shows footing forms under construction. Identify the items specified.

  A. _____

  B. _____

  C. _____

  D. _____

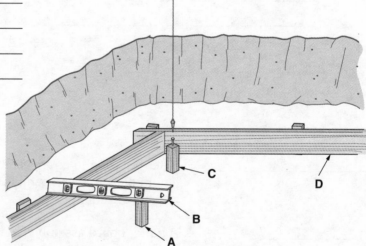

Name _____

10. The drawing shows the lower section of a concrete wall form more than 4′ high. Identify the various parts of the form as indicated.

A. _____

B. _____

C. _____

D. _____

_____ 11. In regular residential construction, basement window openings are usually located level with the top of the ____.

_____ 12. In preparing concrete, the cement and aggregate are mixed together and then water is added. The water causes a chemical action. This action is called ____.

_____ 13. A standard Number 5 reinforcing bar has a diameter of ____″.
   A. 1/4
   B. 3/8
   C. 1/2
   D. 5/8

_____ 14. A standard bag of portland cement contains 1 cu. ft. and weighs ____ lb.
   A. 60
   B. 72
   C. 84
   D. 94

_____ 15. For general-purpose work, concrete can be specified by listing the proportions of sand, cement, and gravel or crushed stone. Stone is seldom over ____" in diameter.
   A. 1/2
   B. 3/4
   C. 1
   D. 1 1/2

_____ 16. Materials added to concrete or mortar to change its properties (such as freezing point or curing time) are called ____.
   A. inhibitors
   B. admixtures
   C. entraining agents
   D. accelerators

_____ 17. The view below shows the installation of a(n) ____.

18. Name the parts of a concrete block indicated in the drawing below.

   A. _____

   B. _____

   C. _____

   D. _____

   E. _____

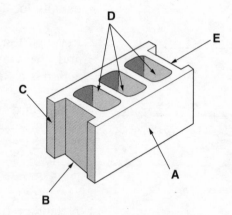

_____ 19. Standard concrete blocks are normally laid with a mortar joint that is ____" thick.
   A. 1/4
   B. 3/8
   C. 1/2
   D. 5/8

Name _____

20. Give the correct name for each type of concrete block shown below.

A. _____

B. _____

C. _____

D. _____

E. _____

F. _____

G. _____

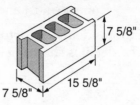

A

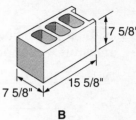

B

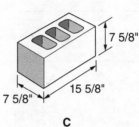

C

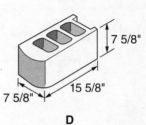

D

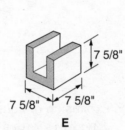

E

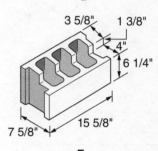

F

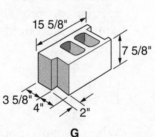

G

_____ 21. An all-weather wood foundation system is constructed from lumber that has been ____.
A. cut to oversized dimensions
B. painted with a special green stain
C. pressure-treated with chemicals
D. exposed to local weather conditions

_____ 22. To ensure good drainage around a wooden foundation, a special basin for water collection is constructed. It is commonly referred to as a ____.
A. storm sewer
B. drainage basin
C. sump pit
D. collector pit

_____   23. It is important that inward forces against a wood foundation be transferred to the floor frame. At places where joists run parallel to the wall, framing members called _____ should be installed between the outside joist and the first interior joist.

24. Slab-on-grade floors for residential structures should be insulated and protected against moisture. An approved detail is shown below. Identify the parts and materials.

A. _____

B. _____

C. _____

D. _____

E. _____

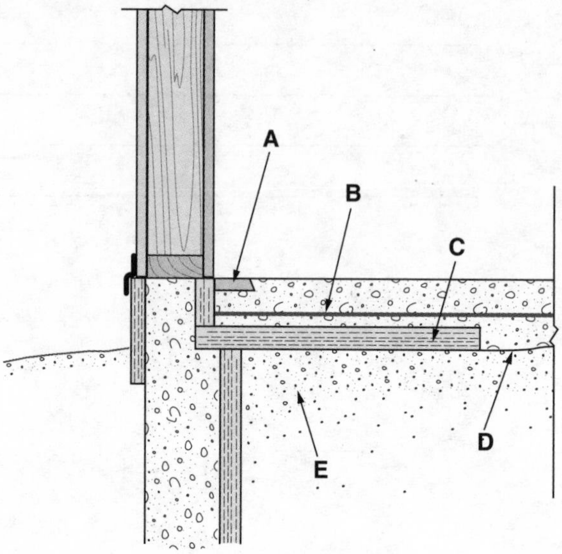

_____   25. Expansion or control joints can be cut in concrete sidewalks or driveways with a power saw equipped with a masonry blade. For regular concrete, this operation should not be carried out until the concrete has cured for at least _____.
   A. 18 hours
   B. 36 hours
   C. two days
   D. three days

_____   26. When placing (pouring) concrete, special protection should be provided when temperatures fall below _____°F.
   A. 0
   B. 20
   C. 32
   D. 40

Name _____

27. How many cubic yards of concrete are needed to pour a basement floor that measures 24′ × 54′ and is required to be 4″ thick? Show your calculations in the space below.

28. How many cubic yards of concrete are required to pour 216′ of footings with a cross section of 8″ thickness × 16″ width? Add 5% for waste and variation in forms. Round your answer up to the nearest 1/4 cu. yd. Show your calculations in the space below.

29. How many cubic yards of concrete are needed to pour a foundation wall 10″ thick, 9′ high, and 20′ long? Do not figure any variation or waste. Round your answer up to the next higher 1/3 cu. yd. Show your calculations in the space below.

30. How many 8″ × 8″ × 16″ concrete blocks are required to lay a foundation wall with a total perimeter of 216′ and specified to be 12 courses high? Show your calculations in the space below.

_____ 31. One method of estimating the number of 8″ × 8″ × 16″ concrete blocks needed for a wall is to divide the square footage of the face area by 100 and then multiply by ____.

CHAPTER **10**

Floor Framing

Carefully study the chapter and then answer the following questions.

__floor flame__   1. Installing the ____ before backfilling helps the foundation withstand the pressure placed on it by the soil.

2. Identify the structural members specified in the drawings below.

    A. __Rim Joist__

    B. __subfloor__

    C. __Floor Joist__

    D. __Sole plate__

    E. __Sill plate__

    F. __Girder__

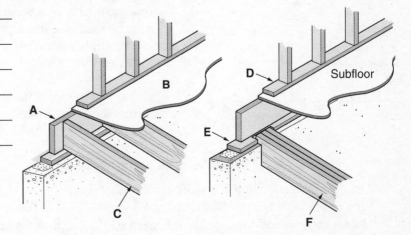

__Western framing__   3. The two basic types of framing are platform and balloon. Platform framing is also referred to as ____ framing.

__ribbon__   4. In balloon framing, the horizontal member that is attached to the studs and carries the second floor joists is called a ____.
    A. plate
    B. sill
    C. ribbon
    D. header

__40__   5. When calculating the size of girders and joists in residential construction, the first floor live load is usually specified as ____ pounds per square foot.
    A. 20
    B. 30
    C. 40
    D. 50

__W-beam (wide flange)__   6. The ____ steel beam is generally used in residential construction.

7. Provide the correct name for each of the parts specified in the drawings below.

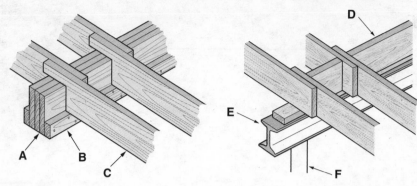

A. ___Girder___
B. ___Ledger___
C. ___Joist___

D. ___bridging block___
E. ___Steel beam___
F. ___Post___

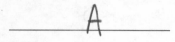

___A___   8. After girders or beams are set in place, the first step in building the floor frame is to ____.
A. lay out the position of the floor joist along the foundation wall
B. lay out the joist spacing along the girders or beams
C. attach the sill to foundation walls
D. cut joist headers to length and set them in place

___All of the above___   9. LVL is often chosen over built-up wood girders because ____.
A. it is dimensionally stable.
B. it is available in lengths to span most houses.
C. it has consistent strength properties
D. All of the above.

10. Building codes usually specify that deflection (bending downward at the center) in a floor joist for residential buildings should not exceed 1/360 of the span under normal loads. What fraction of an inch would this equal for a span of 10′-0″? Show your calculations in the space below.

$$10' = 120'' \ (10 \times 12)$$

$$\frac{1}{360} \times \frac{120''}{1} = \frac{120}{360} = \frac{1}{3}'' \text{ for } 10' \text{ span}$$

Name _____

11. Show the correct distance in inches and fractions of an inch for the measurements identified in the drawing below.

A. _____ 16" _____

B. _____ 16" _____

C. _____ 15 1/4" _____

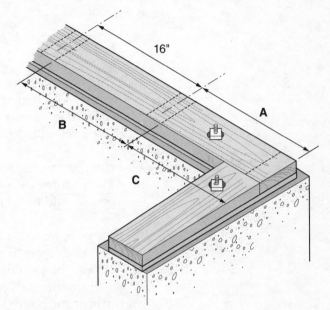

three 16d

12. Floor joists are attached to headers by nailing through the header and into the joists. The text suggests that the nailing pattern consist of _____ nails.

    A. two 20d

    B. three 16d

    C. three 20d

    D. four 16d

rim boards

13. Because I-joists do not expand and contract with changes in moisture content, it is necessary to use engineered _____ in place of solid wood joist headers when framing with I-joists.

14. The drawing below shows the floor framing around an opening. Write the name for each of the specified parts.

A. _____ rim joist _____

B. _____ Regular joist _____

C. _____ tail joist _____

D. _____ Double header _____

E. _____ Double trimmer _____

F. _____ tail joist _____

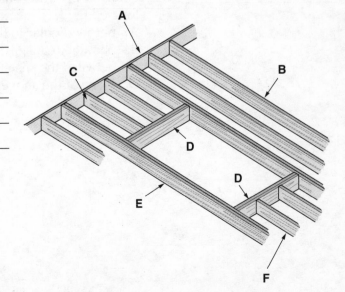

_____12'-0"_____ 15. The National Forest Products Association recommends the use of joist hangers or ledger strips to support tail joists at an opening whenever they span a distance of ____ or more.
A. 8'-0"
B. 9'-0"
C. 10'-0"
D. 12'-0"

_____A_____ 16. Which of the following is *not* a purpose of bridging?
A. Make each joist stronger.
B. Hold joists in a vertical position.
C. Transfer load from one joist to adjacent ones.

_____lowered_____ 17. When certain areas of a floor frame must support extra weight (bathroom fixtures, supporting partitions, or mechanical equipment) the joists should be spaced closer together and/or ____.

_____B_____ 18. If large holes must be cut into a solid wood joist to accommodate plumbing lines, they should be positioned ____.
A. near the top edge
B. approximately in the middle
C. near the bottom edge

_____D_____ 19. Which of the following is *not* permitted when cutting holes in I-joists?
A. Removing more than one of the manufacturer's knockouts.
B. Cutting more than one hole in the web.
C. Holes 1 1/2" in diameter or larger.
D. Holes within 1" of a bearing surface.

_____1"_____ 20. When installing OSB subflooring, fasteners should penetrate the wood joist every ____".
A. 1
B. 3
C. 5
D. 6

_____A_____ 21. OSB and plywood subflooring panels should be installed with the long edge ____.
A. perpendicular to the joists
B. parallel to the joists
C. between the joists
D. None of the above.

Name _____

22. Make an accurate estimate of the number and length of joists and headers required to construct the floor frame for a single-story rectangular building. The joists will be 12′ long and the headers will run 32′ along each of the two walls. Use 2 × 10 lumber spaced 16″ on center. Show your calculations in the space below.

Headers: _____6_____

Joists: _____24_____

Total Board Feet _____647 ft_____

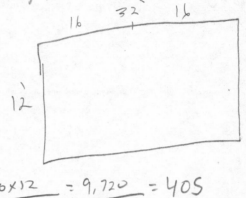

$$\frac{27 \times 2 \times 10 \times 12}{12} = 540$$

$$\frac{4 \times 2 \times 10 \times 16}{12} = 107$$

No. of Joists = Length of wall × 3/4 + 1 + extras

$$\begin{array}{c}32\\12\\\hline\end{array}$$
384/16 = 24

$$\begin{array}{c}32\\ \times .75\\\hline 24\end{array}$$

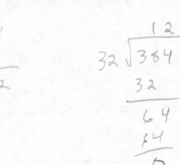

$$\frac{4 \times 3 \times 10 \times 16}{2 \times 12} = \frac{1920}{24} = 80$$

$$\frac{27 \times 3 \times 16 \times 12}{2 \times 12} = \frac{9,720}{24} = 405$$

$$\begin{array}{c}405\\ 80\\\hline 485\end{array}$$

1 1/2
3/2

23. Estimate the number of 5/8 × 4 × 8 sheets of OSB subflooring required for a 12′ × 32′ floor frame. Show your calculations in the space below.

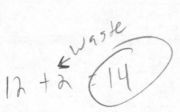

$$\begin{array}{c}12\\ \times 32\\\hline 24\\ 360\\\hline 384\end{array}$$

$$\begin{array}{c}8\\ \times 4\\\hline 32\end{array}$$

$$32\overline{)384}$$ ... = 12

12 sheets of OSB

12 + 2 (waste) = 14

24. Determine the total number of $2 \times 10 \times 16'$ pieces needed to construct the floor frame drawing below. To simplify estimating, ignore the bumped-in portion of the front wall. Show your calculations in the space below.

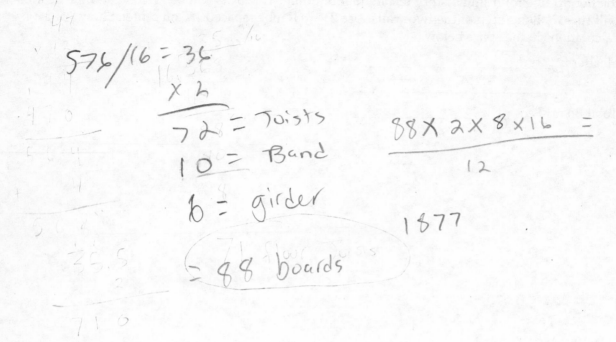

576/16 = 36

× 2

72 = Joists

10 = Band

6 = girder

= 88 boards

88 × 2 × 8 × 16 =

12

1877

25. Estimate the total lineal feet and board feet of 2×6 material required for the floor frame sill of the residential sketch shown below. Round out a given length to the next higher even dimension. For example, the end walls are 28'-8". Round it off to 30'-0". This particular run would probably be made with one 14' piece and one 16' piece. Show your calculations in the space below.

Total lineal ft.: _____156 ft_____

Total bd. ft.: _____1877_____

30
12
____
360"

10 × 1 × 6 × 16
_____
2 × 12

1
48
12
___
96
480
____
576"   24 = 40

96
960

47'-4"   3

2×8 JOISTS
16" O.C.

2×8 JOISTS
16" O.C.

28'-8"   2

2

2'-0"

20'-0"   6'-8"   20'-8"   3

Name _____

26. Determine the number of sheets of OSB or plywood needed to lay the subfloor of a building that measures 24′ wide and 64′ long. What is the total square footage of these pieces? Show your calculations in the space below.

No. of pieces: _____48_____

Total sq. ft.: _____1,536_____

27. Explain what you would do to frame a section of flooring to receive a concrete base so that the finished floor is the same level as adjacent floors.

Frame down lower than the adjacent floor.

_____

Name _____ Date _____ Score _____

CHAPTER 11
Wall and Ceiling Framing

Carefully study the chapter and then answer the following questions.

_____ D _____ 1. In platform construction, wall-framing members include sole plates, top plates, studs, headers, and ____.
A. beams
B. joists
C. sheathing
D. sills

_____ 16" _____ 2. Although studs are sometimes spaced 24" O.C. in residential structures, a spacing of ____" O.C. is more commonly used.
A. 12
B. 16
C. 18
D. 20

3. Study the drawing of a typical wall frame shown below and identify the specified parts.

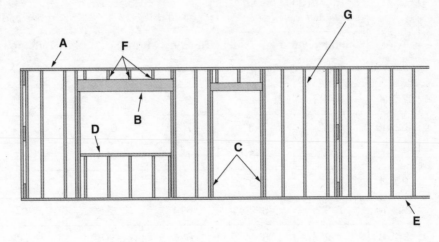

A. _____Top plate_____ E. _____Bottom plate_____
B. _____Header_____ F. _____Cripple studs_____
C. _____Jack Stud_____ G. _____Studs_____
D. _____Sill_____

_____ 4. When assembling studs to form outside corners, it is recommended that ____ nails be used and that they be spaced 12″ apart.
    A. 8d
    B. 10d
    C. 12d
    D. 16d

_____ 5. Where is the best place to look for rough opening sizes when framing exterior walls?
    A. Floor plan.
    B. Building elevations.
    C. Door and window schedule.
    D. Framing plans.

_____ 6. Headers may be formed by nailing two or three members together with a spacer between them. The spacer should be ____″ thick.
    A. 1/4
    B. 3/8
    C. 1/2
    D. 5/8

_____ 7. Refer to the span table below to find the proper size header to support the ceiling and roof with a 50 lb. per square foot snow load for a 5′-4″ rough opening in a house that is 28 feet wide.
    A. 2-2 × 6 or 3-2 × 4
    B. 2-2 × 8 or 3-2 × 6
    C. 2-2 × 10 or 3-2 × 8
    D. 2-2 × 10 or 1-2 × 12

Allowable Header Spans

| Header Supporting | Size | Span with 30 lb per sq. ft. snow load | | Span with 50 lb per sq. ft. snow load | | Span with 70 lb per sq. ft. snow load | |
| | | Building width | | Building width | | Building width | |
		20′	28′	20′	28′	20′	28′
Roof and ceiling	2-2×4	3′-6″	3′-2″	3′-2″	2′-9″	2′-10″	2′-6″
	2-2×10	8′-5″	7′-3″	7′-3″	6′-3″	6′-6″	5′-7″
	3-2×8	8′-4″	7′-5″	7′-5″	6′-5″	6′-8″	5′-9″
Roof, ceiling, and one clear-span floor	2-2×4	3′-1″	2′-9″	2′-9″	2′-5″	2′-7″	2′-3″
	2-2×10	7′-0″	6′-2″	6′-4″	5′-6″	5′-9″	5′-1″
	3-2×8	7′-2″	6′-3″	6′-5″	5′-8″	5′-11″	5′-2″
Roof, ceiling, and two clear-span floors	2-2×4	2′-1″	1′-8″	2′-0″	1′-8″	2′-0″	1′-8″
	2-2×10	4′-9″	4′-1″	4′-8″	4′-0″	4′-7″	4′-0″
	3-2×8	4′-10″	4′-2″	4′-9″	4′-1″	5′-5″	4′-8″

Name _____

_____ B

8. When laying out the sole and top plates for an outside wall section, which of the following steps is *incorrect*?
   A. Lay out all centerlines of door and window openings.
   B. Lay out centerlines of intersecting partitions.
   C. Lay out all regular stud spacings and mark them with an X.
   D. Lay out two stud positions at each side of all openings.

_____ 9. A long measuring stick that represents the actual wall frame with markings made at the proper height for horizontal wall frame members is called a(n) ____.

_____ 10. When laying out stud positions in a rough opening ____.
   A. the stud locations should be adjusted so that a stud falls at the location of one side of the opening
   B. the regular on-center spacing of studs is continued through the opening for cripple studs
   C. cripple stud locations is marked independent of regular stud locations
   D. The carpenter can choose to follow any of the above practices

_____ 11. When preparing to nail studs to plates, they should be laid in position on the subfloor with ____.
   A. the crowns of the studs alternated up and down
   B. the crown of each stud down
   C. the crown of each stud up
   D. Studs with crown cannot be used

12. The drawing below shows a section of a wall frame. Provide the correct nail size normally used at each identified point.

   A. _____
   B. _____
   C. _____
   D. _____

_____ 13. When erecting wall sections, make sure that they are exactly vertical by using either a level and straightedge or a(n) ____.

_____A_____ 14. Studs are sometimes cut to exact length at the mill. These studs are designated by the letters ____.
- Ⓐ P.E.T. _precision end trim_
- B. E.P.T.
- C. T.E.L.
- D. C.T.L.

___backing___ 15. Boards and blocks installed in the wall framing for the sole purpose of mounting plumbing fixtures, towel bars, and other fixtures are called ____.

_____ 16. When let-in wood bracing is required in a wall frame, it is usually made from ____ material.
- A. 1 × 2
- B. 1 × 4
- C. 2 × 2
- D. 1 × 6

_____B_____ 17. When installing the upper half of a double plate, use 10d nails and stagger the pattern with a spacing of ____".
- A. 12
- Ⓑ 16
- C. 20
- D. 24

_____D_____ 18. Joints in the upper top plate should be located at least ____' from those in the lower top plate.
- A. 1
- B. 2
- C. 3
- Ⓓ 4

_____6"_____ 19. OSB and plywood sheathing should be nailed every ____" along the edge.

_____2"_____ 20. The drawing below shows the installation of metal strap bracing for 2 × 4 stud framing. The strap is usually ____ wide.

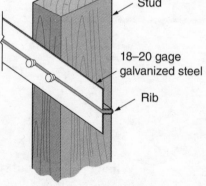

Stud

18–20 gage galvanized steel

Rib

Name _____

Plywood / OSB   21. ____ panels are preferred over foam panels for sheathing because the structural panels brace the wall.

gypsum   22. Where fire resistance is an important consideration, ____ sheathing is often used.
A. gypsum
B. OSB
C. plywood
D. isocyanate foam

Ceiling Joists   23. Main ceiling framing members are called ____.

_D_   24. Gypsum sheathing can be applied with:
A. screws
B. nails
C. adhesive
D. Both A and B.

25. Look at the following illustration. Study the framing of the opening in the partition. Indicate under what conditions the door opening in an interior partition can be framed as shown.

Under the condition of being a non-load bearing wall.

26. After adding the length of all walls and partitions, a carpenter finds there are a total of 270'. How many lineal feet of 2 × 4 stock is needed to build the sole and double plates? How many bd. ft. does this equal? Show your calculations in the space below.

Lineal ft.: _____810 ft_____

Bd. ft.: _____540 ft_____

$$\begin{array}{r} 270 \\ 3 \\ \hline 810 \end{array}$$

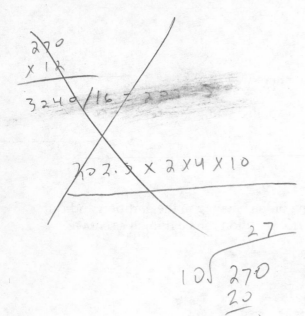

$$\frac{81 \times 2 \times 4 \times 10}{12} = 540$$

$$\begin{array}{r} 27 \\ 10\overline{)270} \\ 20 \\ \hline 70 \end{array}$$

_____202.5_____

27. How many studs are required for the wall framing described in the previous problem? Assume that the studs are spaced 16" O.C. and that there are 14 corners, 13 intersections, and 24 openings. Use the first (and longer) method described in the text. Show your calculations in the space below.

$$\begin{array}{r} 270 \\ \times 12 \\ \hline 3240/16 \end{array} = 202.5$$

Name _____

28. After calculating the total exterior wall surface and subtracting for the major openings, it is found that the net area to be sheathed is 1220 square feet. How many pieces of 4×8 sheathing are required if no allowance is made for waste? Show your calculations in the space below.

$$32\sqrt{1220} \quad = \quad 38\ ^{1}/_{8}$$

29. Describe *structural insulated panels (SIPs)*.

   SIPS are panels made with OSB or plywood sheathing
   on the exterior side, an insulating foam core,
   a structural panel, and usually OSB on the inside.

   House Wrap   30. A thin, tough plastic sheet material applied to exterior walls is called
                             ___.

Name _____ Date_____ Score _____

CHAPTER **12**

Roof Framing

Carefully study the chapter and then answer the following questions.

1. Five basic types of roof designs are shown in the illustration below. Identify each type.

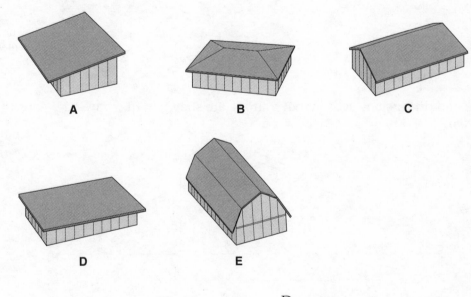

A. _____ D. _____

B. _____ E. _____

C. _____

_____ 2. A traditional roof design, similar to the hip roof except that the sides have a double slope, is called a(n) ____ roof.

_____ 3. The kind of rafters used on both gable and hip roofs include common, hip jack, and ____.
A. cripple jack
B. valley
C. hip
D. valley jack

4. Identify the specified parts of the common rafter shown below.

A. _____

B. _____

C. _____

D. _____

E. _____

F. _____

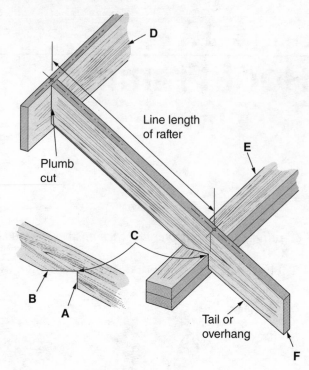

5. Basic terms and dimensions used in roof framing are shown in the drawing below. Identify each specified item.

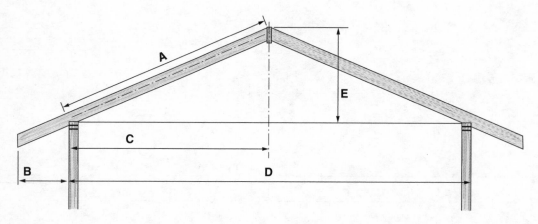

A. _____   D. _____

B. _____   E. _____

C. _____

_____   6. The slope of a roof is sometimes expressed as a fraction formed by placing the _____ over the run.

_____   7. The blade of a standard framing square is _____ inches long.

_____   8. The side of a framing square that shows the manufacturer's name is called the _____.

Name _____

_____

9. Refer to the illustration of a framing square to find the line length of a common rafter with 5-in-12 slope and run of 14'-0".
   A. 70"
   B. 14'-0"
   C. 14'-9 1/8"
   D. 15'-2"

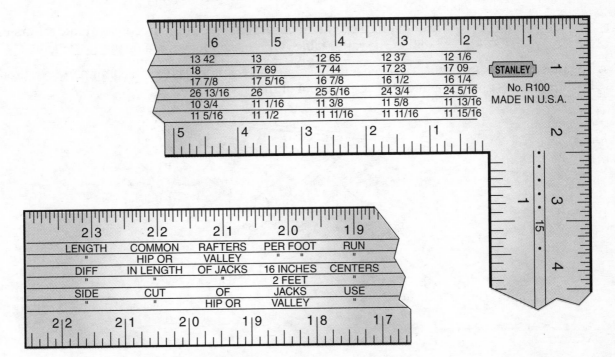

10. Refer to the above illustration of a framing square to find the line length of a common rafter with 4-in-12 slope and run of 8'-6". Round your answer to the nearest 1/8". Show your calculations in the space below.

_____   11. When laying out a common rafter using the step-off method, the second step procedure consists of ____.
    A. shortening the rafter at the ridge
    B. laying out the bird's mouth
    C. setting the rise and run
    D. laying out odd units

_____   12. To use a speed square, the carpenter must know the roof ____.

_____   13. In the step-off method, carpenters laying out conventional rafters use either the rafter tables on the framing square or the ____ method.

14. The illustration below shows a speed square set to make a plumb cut on a common rafter. From the illustration, give the amount of rise per foot of run.

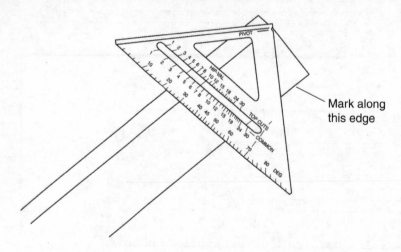

Mark along this edge

_____   15. Which of the following items would not be used to calculate the length of the ridge for a hip roof constructed on a simple rectangular building?
    A. Roof slope.
    B. Roof run.
    C. Length of building.
    D. Thickness of rafter stock.

_____   16. When laying out a hip rafter, the same procedure is used as for the common rafter. An important difference, however, is that ____" is used on the blade of the framing square instead of 12".
    A. 14
    B. 16
    C. 17
    D. 20

Name _____

17. Find the *line length* for a hip rafter when the slope of the roof is 4-in-12 and the run is 10′-2″. The framing square table number is 17.44. Round to the nearest 1/4″. Show your calculations in the space below.

_____ 18. A gable end frame that extends outward over brick veneer can be formed by using lookouts and blocking attached to a ____.
    A. sill
    B. ledger
    C. joist
    D. rafter

_____ 19. The drawing below shows framing for an intersecting hip roof before jack rafters are installed. The framing members shown include ridges, common rafters, hip rafters, and one ____.

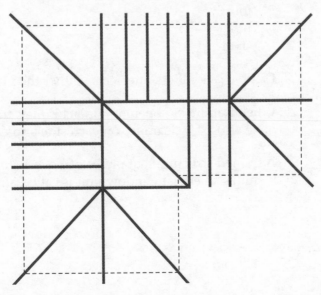

_____ 20. Hip and valley rafters must be shortened at the ridge by a horizontal
distance equal to _____.
A. one-half of the ridge thickness
B. one-half of the 45° thickness of the rafter
C. one-half of the 45° thickness of the ridge
D. one-half of the rafter thickness

21. The common difference in length of a set of jack rafters can be determined by using the framing square as shown in the drawing below. Identify the names of the points and distances specified.

A. _____

B. _____

C. _____

_____ 22. The layout of the bird's mouth and overhang for a hip jack rafter is
made or secured from the _____.
A. common rafter pattern
B. hip rafter pattern
C. framing square using the common difference
D. framing square table

_____ 23. Rafter tables can be used to lay out the side cut of jack rafters. Figures
are secured from the _____ line from the top.
A. third
B. fourth
C. fifth
D. sixth

_____ 24. Cripple jack rafters intersects neither the plate nor the _____.

_____ 25. When framing an opening for a chimney, framing members must
have a(n) _____ clearance on each side and end.

_____ 26. A framed structure that projects above a sloping roof and usually
includes a vertical window unit is called a(n) _____.

Name _____

_____ 27. The drawing below shows horizontal framing members running between rafters on opposite sides of the roof frame. They provide a bracing effect and are called _____.

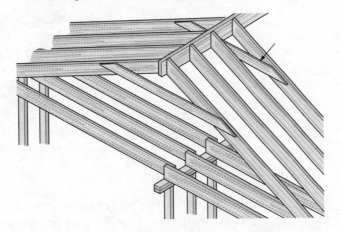

_____ 28. In flat roof framing, the main supporting members are called _____.

29. A standard W truss is shown in the drawing below. Identify the specified members.

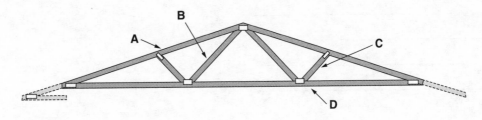

A. _____   C. _____

B. _____   D. _____

_____ 30. A standard roof truss with a length of 24′ usually requires about _____″ of camber.
    A. 1/4
    B. 3/8
    C. 1/2
    D. 3/4

31. Two special truss designs are shown below. Provide the correct name for each.

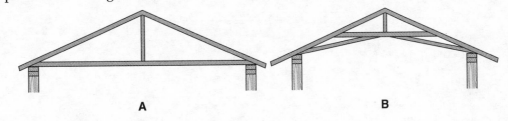

A. _____   B. _____

32. The illustration below shows special, gable-end framing. Name the parts.

A. _____

B. _____

C. _____

D. _____

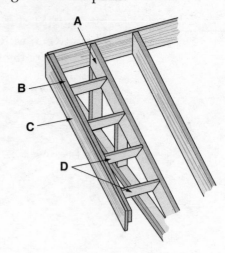

_____ 33. Sheathing provides added strength and rigidity to the roof frame and also serves as a(n) _____ for the roof covering materials.

34. Look at the drawing below and identify the parts of the roof framing.

A. _____

B. _____

C. _____

D. _____

E. _____

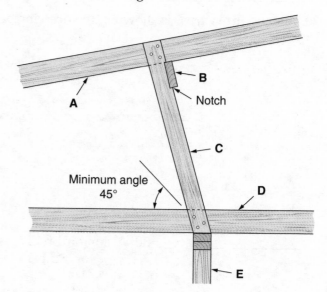

Name _____

35. How many rafters are required for a plain gable roof constructed on a rectangular building measuring 24′ × 48′? The specified rafter spacing is 16″ on center. Show your calculations in the space below.

36. Determine the number and length of 2 × 6 rafter stock needed to frame a gable roof on a 24′ × 42′ building. The slope of the roof is 4-in-12 with a 1′ × 0″ overhang. The rafter spacing is specified as 24″ on center. Show your calculations in the space below.

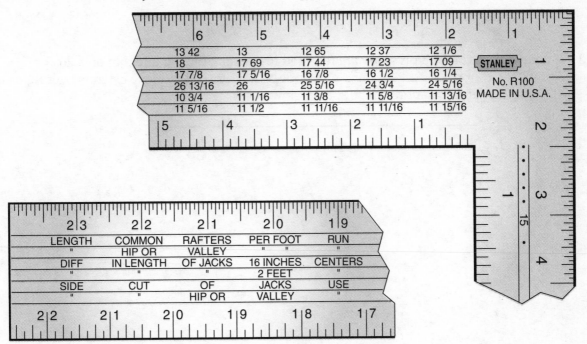

   A. Length of rafter stock: _____

   B. Number of rafters: _____

   C. Total board feet: _____

Refer to the rafter table shown above for problems 37, 38, and 39.

37. Estimate the number of pieces of standard length rafter stock needed to frame a hip roof on the building described in Problem 37. Show your calculations in the space below.

A. Total pieces of 14′ rafter stock: _____

B. Length of stock needed for hip rafters: _____

38. How many pieces of 4 × 8 plywood are required to sheath a plain gable roof on a 26′ × 54′ building? The roof has a 2′ overhang and a slope of 3-in-12, and the rake extends over the gable end 6″. Show your calculations in the space below.

A. No. of pieces: _____

B. Total sq. ft.: _____

Name _____

39. How many pieces of 4 × 8 plywood are required to sheath a flat roof on a 30′ × 66′ building? The roof overhang is 2′-6″ on all sides. Show your calculations in the space below.

    A. No. of pieces: _____

    B. Total sq. ft.: _____

CHAPTER **13**

Framing with Steel

Carefully study the chapter and then answer the following questions.

1. What do each of the lettered parts mean in the Steel Stud Manufactures Association designation shown below.

   A. _____

   B. _____

   C. _____

   D. _____

2. What is an important safety precaution that should be taken when handling light-gauge steel framing?

   _____

_____ 3. What are the advantages of steel framing (*select all correct answers*)?

   A. It is noncombustible.

   B. It is not damaged by insects.

   C. It is less expensive than wood framing.

   D. It does not support mold or fungus.

_____ 4. Sometimes, steel studs may not be used on outside walls because they are poor ____.

5. Name the type of framing detailed below.

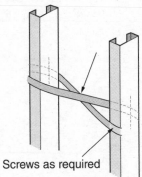

Screws as required

_____

_____ 6. Steel framing contains at least ____% recycled materials.

_____ 7. Most manufacturers of steel framing use a(n) ____ to prevent accidental mixing of different gages.

_____ 8. Steel roof trusses are (*select all correct answers*):
    A. not allowed by most building codes.
    B. sometimes engineered by specialty companies.
    C. sometimes built on-site to code specifications.
    D. required to follow engineering specifications.

9. List the tools usually required when working with steel framing.

_____

_____

10. List three methods of permanent bracing of steel wall frames to prevent racking.

_____

_____

_____ 11. Studs, joists, and ____ are manufactured by brake forming and punching galvanized coil and sheet stock.

_____ 12. Wall covering materials, such as drywall are usually fastened to steel wall framing with ____.
    A. self-tapping drywall screws
    B. ring-shank drywall nails
    C. powder-actuated tool
    D. All of the above are commonly used.

_____ 13. Track for steel-framed walls is usually fastened to concrete floors using ____.
    A. self-tapping drywall screws
    B. ring-shank drywall nails
    C. powder-actuated tools
    D. None of the above.

_____ 14. To resist high-wind forces, plywood should be applied with the long edges ____ to the studs and the ends extending from the top track to the bottom track.

Name _____ **Date** _____ **Score** _____

CHAPTER **14**

Roofing Materials and Methods

Carefully study the chapter and then answer the following questions.

1. List the different materials used to cover a pitched roof.

   _____

   _____

2. If an attic is 40′ × 32′, how much attic venting should be provided? Show your calculation in the space provided.

_____ 3. Roofing materials are estimated and sold by the ____, which is the amount of material needed to provide 100 sq. ft. of finished roof surface.

4. Important distances used in the application of roofing materials are shown below. Provide the correct terms as specified.

   A. _____

   B. _____

   C. _____

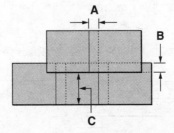
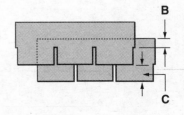

5. The drawing below shows the application of underlayment and metal drip edge. Identify these two materials and the minimum lap distances specified.

A. _____

B. _____

C. _____

D. _____

Sheathing

_____ 6. An ice-and-water barrier, recommended for cold climates, is laid down at the eaves and should extend ____′ inside of the wall line to prevent leak-through from ice dams and wind-blown rain.
A. 1
B. 2
C. 3
D. 6

_____ 7. Shingles are laid along the sides of an open valley so that a 6″ wide waterway starts at the ridge and increases in width by ____″ per foot as they approach the eave.

_____ 8. Short sections of metal flashing are commonly used to waterproof joints between sloping roofs and vertical walls. This type of flashing is usually called ____ flashing.
A. step
B. wall
C. cap
D. corner

_____ 9. To ensure that the top course of shingles will be parallel to the ridge, ____.
A. measure up from the drip edge and snap a chalk line before starting every other course
B. place a long level against the top edge of the shingles as they are laid
C. measure down from the ridge and snap chalk lines as needed
D. measure down from the ridge and snap a chalk line before starting each course

10. How can a starter strip for 3-tab shingles be made with shingles?

_____

Name _____

_____ 11. Three-tab, square-butt shingles, use a minimum of _____ nails per strip.
A. 2
B. 4
C. 5
D. 6

_____ 12. Flashing around a masonry chimney consists of two parts: base flashing that is attached to the roof and _____ or counterflashing that is attached to the chimney.

13. The drawing below shows the applications of hip and ridge shingles with a 5″ exposure. Provide the nailing distances specified.

A. _____

B. _____

_____ 14. For low-sloped roofs, a double underlayment is cemented together to form eave flashing. This flashing should be carried up the roof to a point at least _____″ inside of the interior wall line.
A. 18
B. 24
C. 36
D. 48

_____ 15. Double-coverage roll roofing can be used on slopes as low as _____″ rise per foot.
A. 1
B. 1 1/2
C. 2
D. 2 1/2

_____ 16. Double-coverage roll roofing consists of a granular surfaced area and a smooth area. The smooth area is called a selvage and is _____″ wide.
A. 18
B. 19
C. 20
D. 21

17. The drawing below shows the parts and installation of a flat roof overhang. Identify the specified items.

A. _____

B. _____

C. _____

D. _____

E. _____

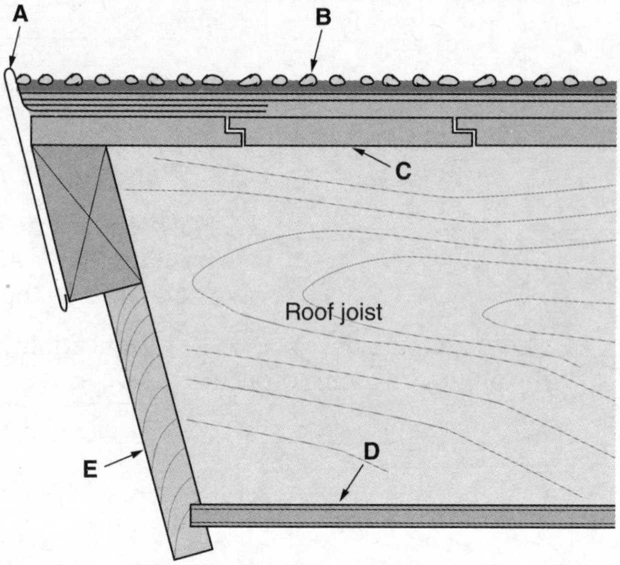

Roof joist

18. Basic construction at the intersection of a flat roof and wall is shown below. Identify the parts and minimum distance as specified.

A. _____

B. _____

C. _____

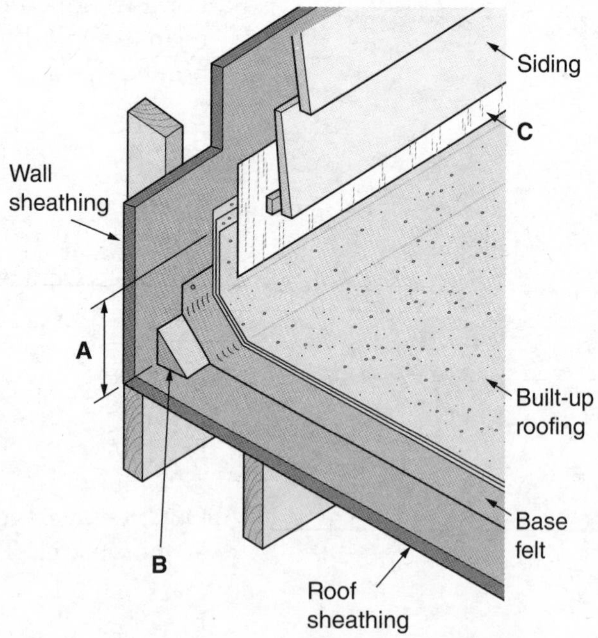

Siding

Wall
sheathing

Built-up
roofing

Base
felt

Roof
sheathing

_____ 19. When applying rubber roofing on a flat or low-pitch roof, allow the rubber membrane to extend ____" up any adjoining walls.
   A. 6
   B. 12
   C. 15
   D. 24

_____ 20. Wood shingles are produced in random widths and in lengths of 16", 18", and 24". They are packaged in bundles. For a standard application, one bundle covers about ____.

Name _____

_____ 21. When using wood shingles on low-sloped (less than 5-in-12) roofs, the exposure should be reduced to a point that provides no less than ____ layers of shingles at any given point.

22. The drawing below shows 18"wood shingles being applied according to standard specifications on a roof with a slope of 6-in-12. Note that each shingle is secured with two nails. Provide the recommended spacing and distances as specified.

Alternate course joints should not align

Gable moulding

Solid wood sheathing

Eave protection

Wood gutter
Fascia
Rafter header

Rafter

First course doubled or tripled

Drip edge 1 1/2"

A. _____   C. _____

B. _____

_____ 23. Wood shakes are highly durable, but must be applied to roofs that have sufficient slope to ensure good drainage. The recommended minimum slope is ____.
A. 4-in-12
B. 5-in-12
C. 5 1/2-in-12
D. 6-in-12

24. Three types of wood shakes are generally available, as shown in the drawing. Identify each type.

A. _____

B. _____

C. _____

A

B

C

_____ 25. Start the wood shakes application by laying down a 36" strip of ____ along the eaves.

_____ 26. Tile roofing, made from concrete or fired clay, weighs from 5.8 to
_____ pounds per square foot.

_____ 27. Strips of wood called _____ are fastened horizontally to the roof deck
to hold tile roofing in place.

_____ 28. In the application of corrugated sheet metal roofing, it is
recommended that side laps be made to include _____ corrugation(s).
A. 1
B. 1 1/4
C. 1 1/2
D. 2

29. Shown below are several components used in a gutter system. Give the general name of each item.

A. _____
B. _____
C. _____
D. _____
E. _____
F. _____
G. _____
H. _____

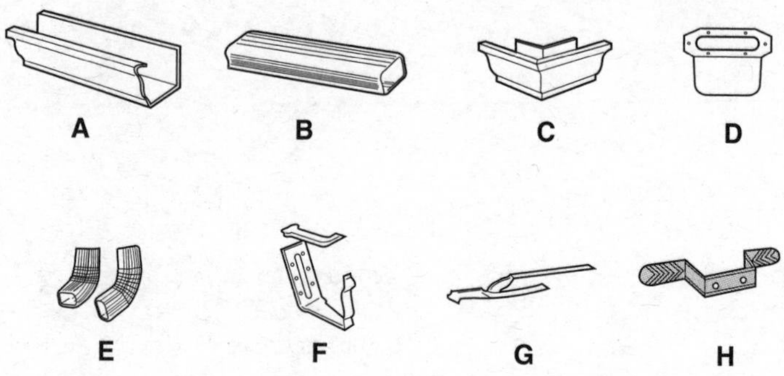

30. The total ground area plus overhang of a house with a gable roof (slope of 5-in-12) is 1880 square
feet. Figure 10% for waste and calculate the number of squares of asphalt shingles required.
Round your answer to the nearest full square. Show your calculations in the space below.

Name _____

31. A rectangular building with a plain gable roof has a common rafter length of 14′. The total length of the ridge is 30′. Calculate the number of full bundles of wood shingles required. Add 10% for waste. Show your calculations in the space below.

32. A 28′ × 48′ building has a hip roof (slope of 6-in-12) with a 2′ overhang. How many full bundles of wood shingles are required? Figure 10% waste and then add one additional square for the hips. Show your calculations in the space below.

33. Estimate the number of squares of asphalt shingles required for a large storage building, 20′ × 80′. The shed type roof (slope of 4-in-12) has a 3′ overhang on the upper edge only. Add 10% for waste. Round your answer to the nearest full square. Show your calculations in the space below.

34. A folding carpenter's rule is being used to determine the pitch of a roof. The reading point on the rule is at 20 1/2. Use the chart below to determine the pitch and slope of the roof.

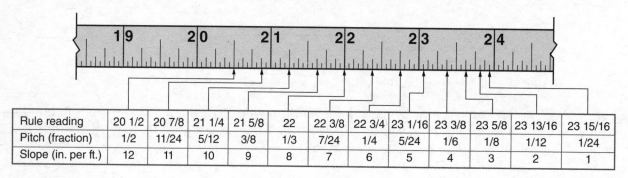

Rule reading	20 1/2	20 7/8	21 1/4	21 5/8	22	22 3/8	22 3/4	23 1/16	23 3/8	23 5/8	23 13/16	23 15/16
Pitch (fraction)	1/2	11/24	5/12	3/8	1/3	7/24	1/4	5/24	1/6	1/8	1/12	1/24
Slope (in. per ft.)	12	11	10	9	8	7	6	5	4	3	2	1

Pitch: _____

Slope: _____

CHAPTER **15**

Windows and Exterior Doors

Carefully study the chapter and then answer the following questions.

_____ 1. Woodworking factories that produce windows and doors are commonly called ____ plants.

_____ 2. Quality control in the manufacturing of windows is based on guidelines established by the ____ Association.
   A. Window and Door Manufacturers
   B. American Architectural Manufacturers
   C. Canadian Standards
   D. All of the above.

_____ 3. A(n) ____ window consists of two sashes that slide up and down in the frame.

_____ 4. When two windows are joined together side-by-side, the portion of the center member, including the parts that are joined is called a ____.
   A. muntin
   B. double jamb
   C. mullion
   D. sister jamb

_____ 5. A ribbon of short window units is sometimes installed high on the wall where it is desirable to ensure privacy or provide wall space for furniture arrangements. ____ windows are especially adaptable to such an installation.

_____ 6. In residential construction, the standard height of windows, measured from the bottom side of the head to the finished floor, is ____.
   A. 6'-6"
   B. 6'-8"
   C. 6'-0"
   D. 7'-0"

_____ 7. Window glass is a major source of heat loss. A single layer of
standard glass has a U-value of about 1.1. A double pane of the same
glass with a 1/2" air space between has an U-value of about ____.
A. 1.6
B. 0.5
C. 1.0
D. 2.1

_____ 8. Which of the following is true of low-emissivity windows?
A. They are less apt to shatter into large pieces when broken.
B. They are tinted.
C. A thin layer of metal oxide is applied to one of the inner surfaces.
D. There is no air space between the panes of glass.

_____ 9. Double and triple glazing improves insulating values and reduces
noise transmission. For a standard, movable sash, two or three layers
of 1/8" glass are fused together with a(n) ____" air space between
each layer.
A. 1/8
B. 3/16
C. 3/8
D. 1/2

_____ 10. The horizontal location of window units is shown on the floor plan.
In masonry construction, this dimension is usually given to the ____
of the opening.
A. centerline
B. edge
C. middle

11. Basically, windows consist of glass panels mounted in a sash, which is installed in a frame.
Standard details of the construction are usually shown in manufacturer's literature in three
section views, as shown below. Provide the correct name for each view.

A   B   C

A. _____   C. _____

B. _____

Name _____

_____   12. Which of the following is *not* a standard window dimension?
      A. Glass size.
      B. Frame thickness.
      C. Rough opening.
      D. Masonry opening.

_____   13. Standard window units can be adjusted to walls of various thicknesses. This adjustment consists of attaching a special piece to the window edges. This piece is called a ____.
      A. sill spreader
      B. jamb extension
      C. jamb ribbon
      D. frame spacer

_____   14. After the window unit is centered in the rough opening and temporarily secured, the next step is to ____.
      A. check the corners with a framing square
      B. nail through the lower end of the side casing
      C. remove any temporary bracing that was installed at the factory
      D. set wedges under the sill to level the unit

_____   15. The actual sizes of glass blocks are somewhat smaller than the nominal sizes. The actual dimensions of a 12″ × 12″ unit are ____.
      A. 3 7/8″ × 11 3/4″ × 11 3/4″
      B. 3 3/4″ × 11 3/4″ × 11 3/4″
      C. 3 5/8″ × 11 5/8″ × 11 5/8″
      D. 3 7/8″ × 11 7/8″ × 11 7/8″

16. What size of opening is required for a glass block installation that is 9 units wide and 11 units high consisting of 6″ blocks? Show your calculations in the space below.

Width: _____

Height: _____

_____   17. When replacing old windows, one of the first steps is to remove the _____ and lift out the lower the sash.

_____   18. Skylights that consist of a hinged sash should not be installed in roofs with less than a _____ slope.
      A. 5-in-12
      B. 4-in-12
      C. 3-in-12
      D. 2-in-12

_____   19. Exterior doors for residential construction are usually _____ high.

20. Exterior door frames are similar to window frames. Section views show details and sizes. Identify the detailed views shown below.

    A. _____

    B. _____

    C. _____

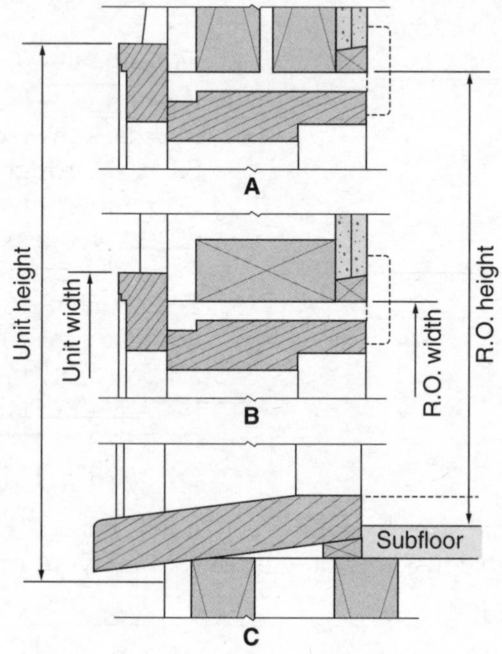

_____   21. When installing an exterior door frame, extra wedges or blocking should be located in the approximate position of the lock strike plate and the _____.

_____   22. The IRC specifies that a doorsill should be _____.
      A. no more than 1/2″ above the finished floor
      B. flush with the finished floor
      C. no more than 1/2″ below the finished floor
      D. The IRC does not specify the height of doorsills.

_____   23. Prehung exterior door units are fastened in place by driving _____ casing nails through the jambs, shims, and into the structural frame members.
      A. 8d
      B. 10d
      C. 12d
      D. 16d

Name _____

_____ 24. When installing a sliding door frame, it is generally recommended that ____ before setting the frame into place.
   A. a layer of the subfloor be removed
   B. a bead of sealing compound be applied
   C. one door be set in place and checked
   D. one side jamb be plumbed and nailed

_____ 25. There are three basic types of garage doors: hinged or swinging, ____, and overhead or roll-up.

_____ 26. Standard heights for residential garage doors include 7'-0" and ____.
   A. 6'-8"
   B. 6'-10"
   C. 7'-6"
   D. 8'-0"

_____ 27. To offset the weight of roll-up garage doors, an effective counterbalance is required. The two most commonly used in residential installations are the extension spring and the ____ spring.

Name _____   Date_____   Score _____

CHAPTER **16**

Exterior Wall Finish

Carefully study the chapter and then answer the following questions.

_____   1. Exterior ____ includes the application of all exterior surfaces of a structure, including the roofing materials. It includes the construction of the cornice and rake and the application of siding and trim members around doors and windows.

_____   2. An open cornice is used when the style of architecture requires the ____ to be exposed to view.

3. The drawing below shows the structural and trim members of a normal boxed cornice. Identify the parts specified.

   A. _____

   B. _____

   C. _____

   D. _____

   E. _____

   F. _____

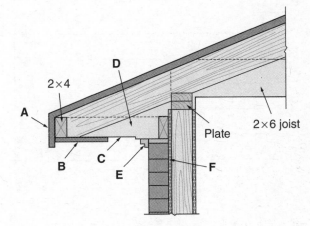

_____   4. When plywood or other thin material is used for the soffit of a boxed cornice, the edge along the fascia should be supported. This can be accomplished with a groove cut in the fascia or by a(n) ____ attached to the rafters and/or fascia.

_____   5. A closed or boxed cornice should have screened slots or ventilating units installed in the ____.

6. The rake is the part of the roof that overhangs a gable. Shown below is a wide box rake section. Identify the trim members as specified.

A. _____

B. _____

C. _____

D. _____

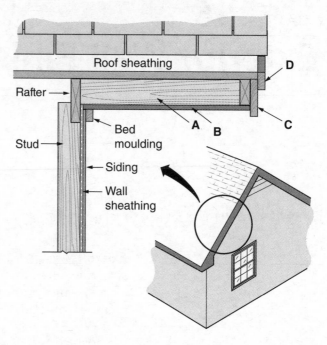

7. Metal soffit material comes in rolls and is generally available in width up to _____".

8. The end views of various types of horizontal siding are shown in the drawing below. Identify each type.

A. _____

B. _____

C. _____

D. _____

E. _____

F. _____

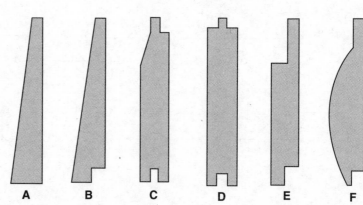

9. When siding is applied over sheathing that consists of solid wood, plywood, or nail-base fiberboard, the nails can be horizontally spaced about _____" apart.
A. 12
B. 16
C. 20
D. 24

Name _____

10. When windows or doors are not protected by a wide overhang, certain extra installations should be made. Identify the items specified in the drawing below.

A. _____

B. _____

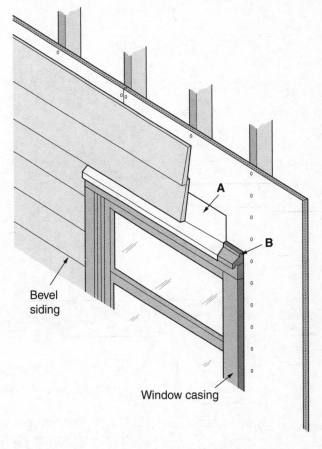

Bevel siding

Window casing

11. The drawing below shows the installation of the first two courses of bevel siding. Identify the specified items.

A. _____

B. _____

C. _____

D. _____

E. _____

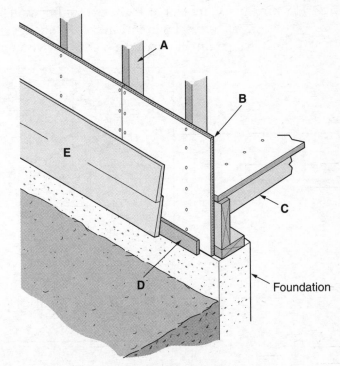

Foundation

_____ 12. Plain beveled siding is lapped so it will provide a tight exterior wall
covering. In a standard installation, 10" beveled siding is lapped
about ____".
A. 1
B. 1 1/2
C. 1 3/4
D. 2

_____ 13. The view below shows a carpenter using a(n) ____ to lay out the position of siding courses at corners and openings.

_____ 14. A(n) ____ is a notched piece of wood that is used to properly position
horizontal siding as it is applied.

_____ 15. When wooden corner boards are used for a horizontal siding application, they are installed ____ (_before, after_) the siding is applied.

_____ 16. Wide bevel siding often has shiplapped or ____ joints.

17. How many square feet of 1 × 8 bevel siding will be required to cover a wall 8'-0" high and
30'-0" long? The total window area is 30 sq. ft. Use the table below and round your answer to the
nearest even number of square feet. Show your calculations in the space below.

Type	Size (inches)	Lap (inches)	Multiply net wall surface by
Bevel siding	1 × 4	3/4	1.45
	*1 × 5	7/8	1.38
	1 × 6	1	1.33
	1 × 8	1 1/4	1.33
	1 × 10	1 1/2	1.29
	1 × 12	1 1/2	1.23
Rustic and drop siding (shiplapped)	1 × 4		1.28
	*1 × 5		1.21
	1 × 6	————	1.19
	1 × 8		1.16
Rustic and drop siding (dressed and matched)	1 × 4		1.23
	*1 × 5		1.18
	1 × 6	————	1.16
	1 × 8		1.14

*Unusual sizes.

Name _____

18. Figure the amount of 1 x 10 bevel siding required to cover a 24′ × 50′ building. The outside walls are 9′ high and the total rise of the gable roof is 4′. Total window and door area is 200 sq. ft. Add 10% to the area of the gable ends for waste. Use the table below and round your answer to the nearest even number of square feet. Show your calculations in the space below.

Type	Size (inches)	Lap (inches)	Multiply net wall surface by
Bevel siding	1 × 4	3/4	1.45
	*1 × 5	7/8	1.38
	1 × 6	1	1.33
	1 × 8	1 1/4	1.33
	1 × 10	1 1/2	1.29
	1 × 12	1 1/2	1.23
Rustic and drop siding (shiplapped)	1 × 4		1.28
	*1 × 5	_____	1.21
	1 × 6		1.19
	1 × 8		1.16
Rustic and drop siding (dressed and matched)	1 × 4		1.23
	*1 × 5	_____	1.18
	1 × 6		1.16
	1 × 8		1.14

*Unusual sizes.

_____ 19. Before applying bevel siding, a narrow spacer strip about ____ should be installed at the bottom of the wall.
    A. the same thickness as the thin edge of the siding
    B. the same thickness as the thick edge of the siding
    C. 1/2″ thick
    D. 3/4″ thick

_____ 20. Battens are applied over vertical siding consisting of solid boards and are attached by nailing ____.
    A. through the center
    B. along one edge
    C. along both edges

_____ 21. Matched vertical siding made from solid lumber should be no more than 8″ wide and installed with two 8d nails spaced no more than ____′ apart.
    A. 2
    B. 4
    C. 6
    D. 8

_____ 22. When applying single-course sidewall shingles up to 10 ″ wide, two nails should be ____ above the butt line of the next course.

_____ 23. Wood shingles are sometimes used for wall coverings. In a single-course application of 16″ shingles, the recommended maximum exposure is ____″.
A. 4 1/2
B. 5 1/2
C. 6 1/2
D. 7 1/2

_____ 24. What is the greatest advantage of shingle panels over regular shingles?
A. They save time.
B. They are less expensive.
C. They are more weather-resistant.
D. All of the above.

_____ 25. When cutting fiber-cement siding with a circular saw, what type of blade should be used?
A. A fine-tooth plywood blade.
B. A special blade for fiber-cement products.
C. A carbide-tipped blade with at least 56 teeth.
D. A fine-tooth blade installed backward.

26. What extra step must be taken behind end joints in fiber-cement siding?

_____

_____ 27. Which of the following is an acceptable method for cutting vinyl siding?
A. Use a circular saw with a fine-tooth blade mounted backward.
B. Use tin snips or aviation snips.
C. Score the surface with a utility knife then break it.
D. All of the above.

_____ 28. When installing a vinyl siding system, allowances must be made for expansion and contraction caused by ____.
A. aging
B. temperature changes
C. moisture

29. What is used to cover the ends of vinyl siding?

_____

_____ 30. The base for stucco consists of sheathing, sheathing paper, and heavily galvanized metal lath. The lath should be spaced ____″ away from the sheathing surface.
A. 1/8
B. 1/4
C. 3/8
D. 7/16

_____ 31. EIFS stands for exterior ____ and finish systems.

Name _____

_____ 32. *True or False?* EIFS wall coverings can be installed only over concrete or concrete block.

_____ 33. To ensure that water or moisture is not trapped behind brick veneer, small holes (called weep holes) are made in the joints of the lowest course of bricks. It is recommended that weep holes be spaced ____" apart.
A. 16
B. 32
C. 40
D. 48

_____ 34. Which one of the following statements is *incorrect* concerning modern shutters?
A. Modern shutters are used mainly for decorative effects.
B. Shutters are usually attached with special, concealed hinges.
C. Shutters are available in standard widths that range from 14" to 20".
D. Shutter installation should permit easy removal for painting and maintenance.

Name _____ Date _____ Score _____

CHAPTER **17**

Thermal and Sound Insulation

Carefully study the chapter and then answer the following questions.

_____ 1. In a normal sequence of residential construction, insulation materials are not installed until the rough-in of plumbing, ____, and electrical wiring has been completed.

_____ 2. Heat is transferred through floors, walls, ceilings, windows, and doors at a rate that varies with the temperature (inside and outside) and the ____ to heat flow provided by the intervening materials.

_____ 3. Heat moves from one molecule to another within a given material or from one material to another when they are in direct contact. This method of heat movement or transmission is called ____.

4. The diagrams below show three methods commonly used to heat enclosed spaces. Provide the correct term used to identify these heat transmission methods.

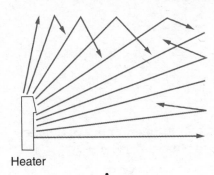

Heater

A

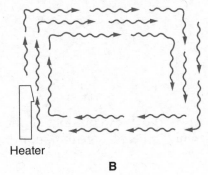

Heater

B

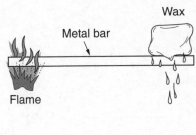

Wax

Metal bar

Flame

C

A. _____   C. _____

B. _____

5. List the various materials used to produce commercial insulation.

_____

_____

_____

_____

_____

_____ 6. The coefficient of thermal conductivity (k) is the amount of heat transferred in one hour through one sq. ft. of a given material that is _____ thick and has a temperature difference of 1°F between its surfaces.

_____ 7. How many British thermal units (Btu) of heat are required to raise 16.5 lb. of water from 33°F to 97°F? Show your calculations in the space below.

_____ 8. A degree day is the product of one day and the number of degrees Fahrenheit the average temperature for that day is below _____°F.

Name _____

_____   9. A U-value is like a k-value, except the structure may consist of
                               several materials, thicknesses, and air spaces. Pictured below is a
                               ceiling section. Select the correct U-value for the insulated area as
                               indicated.
                               A. U = .025
                               B. U = .05
                               C. U = .08
                               D. U = .03

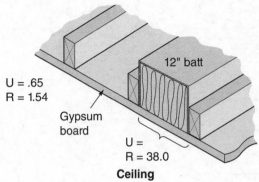

U = .65
R = 1.54

Gypsum
board

12" batt

U =
R = 38.0

Ceiling

10. Calculate the U-value of a wall where the R-values of sheathing, siding, air spaces, plaster, and
    insulation total 11.5. Round your answer to the third decimal place. Show your calculations in
    the space below.

_____   11. When warm, moist air is cooled, some of the moisture is released as
                               condensation. The temperature at which this occurs for a given
                               sample of air is called the ____.

12. Shown below are two wall sections that are identical except for insulation. Select the correct R-value for the insulated wall as specified.
    A. R-10
    B. R-12
    C. R-14
    D. R-19

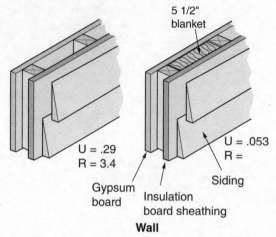

5 1/2"
blanket

U = .29
R = 3.4

U = .053
R =

Gypsum board

Insulation board sheathing

Siding

Wall

13. A wall constructed of 2 × 6 studs is sheathed with 3/4" insulation board, wrapped with Tyvek®, and sided with 3/4" wood bevel siding. Insulation consists of 5 1/2" glass fiber batts over which has been installed 5/8" gypsum board. The approximate R-value of the wall is ____. Show your calculations in the space below.

14. Thermal insulation may be grouped into broad categories. Which of the following is *not* one of those categories?
    A. Loose fill.
    B. Flexible.
    C. Reflective.
    D. Rigid.
    E. All of the above are categories of insulation.

Name _____

15. Three types of insulation commonly used in walls and ceilings of residential structures are shown below. Identify each one.

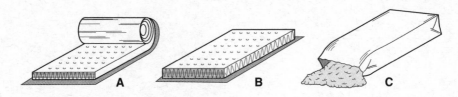

A. _____   C. _____

B. _____

_____   16. ____ is manufactured in sheets and has higher R-values per inch than other forms of insulation.
  A. Batt insulation
  B. Rigid insulation
  C. Loose fill insulation
  D. Blanket insulation

_____   17. Reflective insulation must be exposed to an air space to be effective. The minimum recommended depth of this space is ____".
  A. 1/2
  B. 3/4
  C. 1
  D. 1 1/2

_____   18. The view below shows the essential requirements for an unheated crawl space. Insulation is placed between the joists. The vapor barrier used to cover the ground should be ____ thickness polyethylene film.
  A. 1/16"
  B. 2 mil
  C. 4 mil
  D. 6 mil

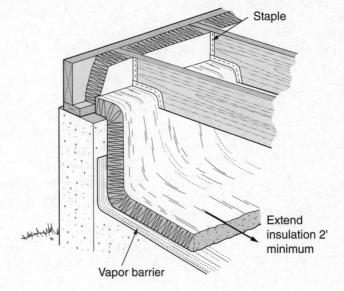

Staple

Extend insulation 2' minimum

Vapor barrier

19. One method of insulating the foundation wall of an existing structure is shown below. Identify the materials and parts as specified.

A. _____

B. _____

C. _____

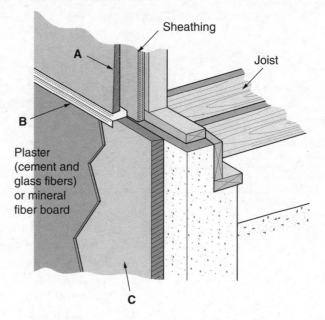

20. The International Residential Code requires ventilation in crawl spaces. How much vent area is required for a 450 square foot crawl space?

_____ 21. A properly installed vapor barrier protects walls, ceiling, and floor from moisture originating within a heated space. To be effective, the vapor barrier in an outside wall should be located ____.
    A. on the outside of the sheathing
    B. between the sheathing and wall frame
    C. between the inside wall covering and wall frame
    D. on the outside surface of the insulation

_____ 22. Many residential structures are built on concrete slab floors that are insulated and contain a vapor barrier. Select the *incorrect* statement concerning this type of construction.
    A. The vapor barrier must be continuous under the entire floor.
    B. Insulation can extend inward under the floor or be carried downward along the foundation.
    C. Both the insulation and vapor barrier should be continuous under the entire floor.
    D. Perimeter insulation is usually extended in from the foundation a horizontal distance of about 2'-0".

Name _____

_____ 23. In a standard gable or hip roof design, soffit vents (as shown below) are commonly used. The total minimum area requirement is expressed as a fraction of the total ceiling area of the structure. The correct inlet figure for the soffit vent shown is _____.
  A. 1/300
  B. 1/500
  C. 1/900
  D. 1/1600

_____ 24. The flanges of blanket insulation are stapled to either the face or side of the framing members. When attaching the flange to the inside edge of a stud, the staples should not be more than _____″ apart.
  A. 8
  B. 12
  C. 10
  D. 16

_____ 25. Perimeter insulation for a concrete slab floor requires a rigid material. A foamed plastic board commonly used is made from _____.
  A. polyethylene
  B. polyurethane
  C. polystyrene
  D. polyester

_____ 26. The installed R-value for fill insulation varies, depending on which installation method, pouring or _____, is used.

27. Estimate the number of square feet of area (walls and ceilings) to be insulated in the house plan shown below. Do not include the ceiling of the garage or its outside walls. Round any inch dimensions to a full foot. Deduct 60 sq. ft. for the sliding glass doors in the family room, but make no other allowances. Round your answers to the next highest 10 square feet. Show calculations in the space below.

Walls: _____

Ceiling: _____

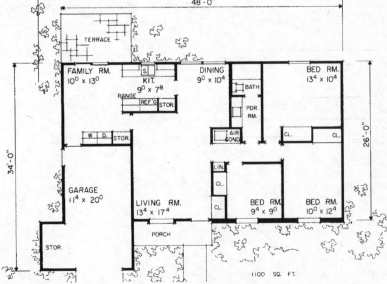

28. List three items of clothing recommended for safe handling of insulation materials.

_____

_____

_____

_____ 29. A blower door test is used to measure ____.
   A. the efficiency of a heating system
   B. the efficiency of a cooling system
   C. leaks in HVAC ducts
   D. air leaks in the exterior surfaces of a building

_____ 30. The unit of measure used to indicate the intensity or loudness of sound is the ____.

_____ 31. Noise is a mixture of sounds, each with a different ____.

_____ 32. As sound moves through any type of wall or other barrier, its intensity is reduced. This reduction is called sound transmission loss (STL). If a given sound of 75 dB is reduced to a level of 36 dB after passing through a wall, the STL rating of the wall is ____.

_____ 33. A system of rating the sound-blocking efficiency of a wall, floor, or ceiling has been established through extensive research. Values (called STC numbers) have been assigned to a wide range of structures and systems. The letters STC stand for ____.
   A. Standard Transmission Coefficient
   B. Systems Transmission Class
   C. Sound Transmission Capability
   D. Sound Transmission Class

_____ 34. Which of the following should have the highest STC?
   A. A partition separating a kitchen from a family room.
   B. Ceilings of bedrooms.
   C. A partition between a bathroom and a bedroom.
   D. A partition separating a bedroom from a hallway.

Name _____

35. The drawings below show section views of four partitions that would be practical to reduce sound transmission in residential construction. Provide the STC rating for each.

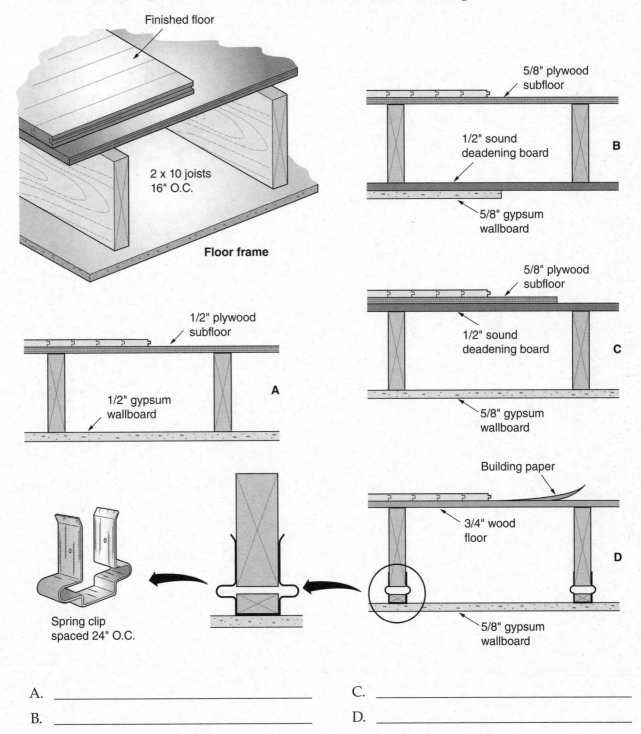

A. _____   C. _____

B. _____   D. _____

_____   36. To secure the best possible soundproofing in a wall structure, consideration must be given to openings for convenience outlets, medicine cabinets, and recessed shelving. For example, convenience outlets and switch boxes on opposite faces of a partition should not be located in the same ____.

_____ 37. A standard, hollow-core, interior door that is carefully fitted has a sound reduction value or transmission loss of about ____.
A. 20–25 dB
B. 25–30 dB
C. 30–35 dB
D. 35–40 dB

_____ 38. If properly done, painting of insulation material will ____ the material's efficiency.
A. increase
B. decrease
C. not affect

CHAPTER **18**

Interior Wall and Ceiling Finish

Carefully study the chapter and then answer the following questions.

_____ 1. Gypsum wallboard is used for standard drywall construction. Panels are _____ wide.

2. Name the edge styles for the drywall illustrations shown below.

A. _____

B. _____

C. _____

D. _____

E. _____

_____ 3. Wallboard panels are often attached with special screws. Since screws hold the wallboard more securely than nails, screw spacing on ceilings can be extended to a maximum of _____ on side walls.

_____ 4. *True or False?* Horizontal joints are harder to treat because they are lower on the walls.

_____ 5. _____ is a fiber-reinforced panel material used as an underlayment for finishing materials on walls, floors, and countertops.

_____   6. In double-layer or two-ply drywall construction, an adhesive is normally used to laminate the finish layer to the base layer. The finish layer is applied so that joints are offset by a distance of at least ____″ from those in the base layer.
A. 8
B. 10
C. 12
D. 16

_____   7. A special gypsum wallboard that will resist mold is easily identified by its ____ facing paper.
A. blue
B. gray
C. green
D. yellow

_____   8. The paper face of gypsum wallboard should not be ____ by the screw head.

_____   9. For installation of corner bead for a single-coat application of veneer plaster, the bead should be set for a ____″ thickness.
A. 1/16
B. 3/32
C. 1/8
D. 3/16

_____   10. A drywall screw gun uses a ____ to control the depth to which screws are driven.
A. clutch
B. depth gage
C. screw stop
D. manual trigger switch

_____   11. *True or False?* Inside corners in drywall applications are reinforced by metal corner beads that are installed before the drywall is applied.

_____   12. Veneer plaster is a high-strength material. It requires a minimum drying time of about ____ hours.
A. 12
B. 24
C. 36
D. 48

_____   13. Gypsum wallboard may be fastened to concrete or masonry walls with ____.
A. adhesive applied to the concrete or masonry
B. furring strips
C. Both A and B.
D. Gypsum wallboard should not be used on concrete or masonry.

_____   14. *True or False?* Drywall compound will begin to cure and get stiff at some time after it has been mixed. Adding a small amount of clean water will extend its workable life.

Name _____

_____ 15. Interior solid wood paneling, at the time of installation, should have about the same moisture content as it will attain after the structure is occupied. For most parts of the country, this is about ____.
A. 6% to 8%
B. 8% to 10%
C. 10% to 12%
D. 12% to 15%

_____ 16. When making a vertical installation of solid wood paneling consisting of nominal 1″ tongue-and-groove boards, nailing or furring strips are required at the top and bottom with intermediate strips spaced no farther than ____″ apart.
A. 24
B. 32
C. 36
D. 48

_____ 17. *True or False?* Solid paneling installed horizontally does not require furring strips.

_____ 18. Plaster requires some kind of a base on which the plaster can be spread. A commonly used base is gypsum lath, which is available in a panel size of ____.
A. 12 × 32
B. 12 × 48
C. 16 × 32
D. 16 × 48

_____ 19. When making an installation of 3/8″ gypsum lath, which of the following specifications is *not* correct?
A. Lath must be nailed at every stud or joist crossing.
B. Use a No. 13 GA nail with a minimum length of 1″.
C. Nails should not be closer than 3/8″ to ends and edges.
D. When using staples, the crown should be parallel to the long dimension of the framing member.

_____ 20. To control plaster coat thickness and provide a level surface or edge, wood or metal strips are attached to doors and other openings. These strips are generally called plaster ____.

21. A gypsum lath installation must be reinforced in certain areas. Identify the reinforcing materials in the drawings below.

A. _____

B. _____

C. _____

_____   22. Gypsum wallboard can be installed over metal or wood _____ attached to a masonry wall.

23. Study the illustration below of a plastered wall section. Name the various elements.

A. _____

B. _____

C. _____

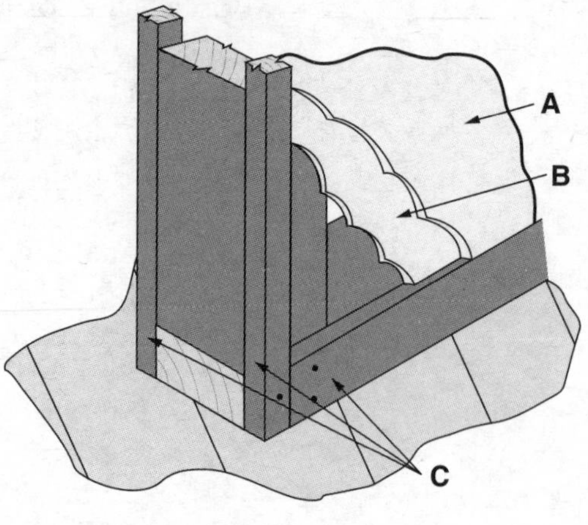

_____   24. In standard three-coat plaster applications, the first coat is called the _____ and is applied directly to the plaster base.

_____   25. In standard plastering applications, which of the following statements is *incorrect*?
   A. In most residential plastering, the first two coats are applied almost simultaneously.
   B. Minimum plaster thickness (including all coats) should not be less than 1/2″ when applied over regular gypsum lath.
   C. When plaster is applied to metal lath, the total thickness measured from the back side of the lath should not be less than 3/4″.
   D. A 1/2″ thickness of plaster has almost twice the resistance to bending or breaking as a 3/8″ thickness.

_____   26. Which of the following statements is *not* correct about cement board?
   A. Does not withstand moisture well.
   B. Can be worked with tungsten carbide cutting/sawing/drilling tools.
   C. Usually consists of a fiberglass-reinforced mix of cement.
   D. Is fireproof and resists impact.

_____   27. Ceiling tiles are often used in remodeling work since they can be applied to nearly any surface. Which one of the following would be of *least* importance in selecting a given product?
   A. Cost.
   B. Unit weight.
   C. Fire resistance.
   D. Light reflection.

Name _____

28. Some ceiling tiles are designed with special tongue-and-groove joints that make them easy to apply with staples. Identify the parts of the joint shown below.

    A. _____

    B. _____

    C. _____

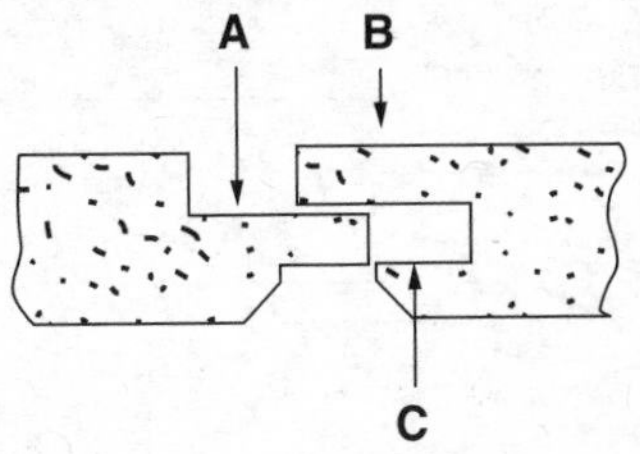

_____ 29. When installing furring strips to exposed joists where one or two of the joists extend below the plane of the others, it is best to ____.
    A. use tapered wedges between joists and strips
    B. plane off the lower edge of the joist
    C. cut notches in lower edge of low joists
    D. use a double layer of furring strips

30. Calculate the amount of 1 × 6 tongue-and-groove solid wood paneling needed for one wall of a room 23′ long with an 8′ height. There are no openings and the boards will be applied vertically. Use the area factor of 1.17 and add 7% for waste. Round off your answer to the next higher full board foot. Show your calculations in the space below.

_____ 31. When installing ceiling tile using adhesive or staples, the best procedure is to start setting the tile ____.
    A. in any one of the corners
    B. along the main centerline of the room
    C. at a midpoint on either wall
    D. at the center of an end wall

32. How many 4 × 8 sheets of plywood are required to panel the wall described in question 30? The panels will be applied vertically and the studs are all equally spaced 16″ on center. Show your calculations in the space below.

No. of panels: _____

Sq. ft. of plywood: _____

33. Determine the amount of standard gypsum lath in sq. ft. for a building with 1140 sq. ft. of floor area and the number of bundles of lath necessary. The total length of all inside wall surfaces is 362′ and the ceiling is 8′ high. Make no allowances for doors and windows and round your answer to the nearest full bundle of lath. Show your calculations in the space below.

Sq. ft.: _____

Name _____

_____ 34. Plasterers base their price estimates on the total number of square yards. Determine the square yards of plaster needed for the structure in question 33. Round your answer to the next higher square yard. Show your calculations in the space below.

35. An 8′ ceiling measures 13′-7″ × 20′-8″. How many standard 12 × 12 tiles are required to cover the ceiling if a balanced pattern (both lengthwise and crosswise) is specified? How many standard cartons must be ordered if no other matching tiles are available? Show your calculations in the space below.

No. of tiles: _____

No. of cartons: _____

CHAPTER **19**

Finish Flooring

Carefully study the chapter and then answer the following questions.

_____   1. _____ flooring is any type of material laid down as the final flooring surface.

_____   2. Hardwood flooring is generally available in thicknesses of 3/8″, 1/2″, and ____″.
  A. 5/16
  B. 7/16
  C. 3/4
  D. 25/32

_____   3. The best grade of plain-sawed oak flooring is designated by the term ____.
  A. first
  B. clear
  C. select
  D. premium

_____   4. When installing standard wood strip flooring over concrete, use 4d casing nails and space them ____″ apart.
  A. 8
  B. 10
  C. 12
  D. 16

_____   5. In the illustration below, what type of tool is being used to install strip flooring?

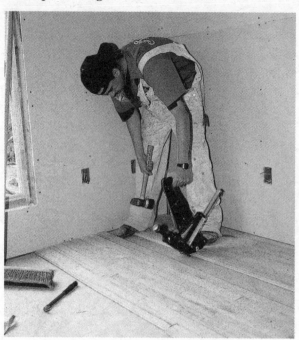

6. A hardwood flooring installation may start along a sidewall, as shown in the following illustration. The first strip is carefully aligned and then nailed. Identify the types of nailing used and provide the recommended spacing and angles.

A. _____

B. _____

C. _____

D. _____

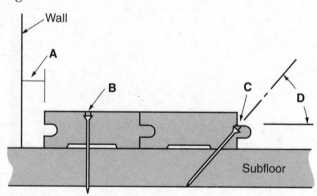

_____   7. When installing strip flooring, it is recommended that end joints be spaced a minimum of ____" apart in adjacent courses.
    A. 6
    B. 8
    C. 12
    D. 16

_____   8. To reverse the direction of laying in a section of the installation, a groove can be converted into a tongue by inserting a hardwood strip. This strip is commonly referred to as a(n) ____.

Name _____

9. Determine the amount of 3/4″ × 3 1/4″ strip flooring required for a 17′ × 28′ room. Refer to the table below. Provide the number of board feet and the number of full bundles required. Show your calculations in the space below.

   Bd. ft.: _____

   Bundles: _____

Flooring board size	Additional percentage
3/4" × 1 1/2"	55%
3/4" × 2"	43%
3/4" × 2 1/4"	38%
3/4" × 3 1/4"	29%
3/8" × 1 1/2"	38%
3/8" × 2"	30%
1/2" × 1 1/2"	38%
1/2" × 2"	30%

10. Calculate the amount of 3/8″ × 2″ strip flooring needed for a living room (14′ × 24′) and entrance hall (4′ × 10′). Provide the total number of board feet required for the two areas and the total number of full bundles that should be ordered. Show your calculations in the space below.

    Bd. ft.: _____

    Bundles: _____

11. Hardwood strip flooring (3/4″) can be installed over a concrete slab. An approved system, where the concrete is in contact with the ground, is shown in the following drawing. The vapor barriers are polyethylene film. Identify these and other items as specified.

    A. _____

    B. _____

    C. _____

    D. _____

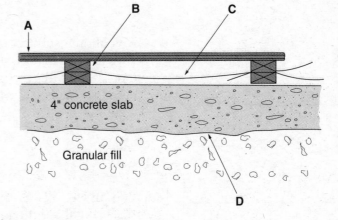

_____ 12. Laying out several rows of flooring at a time and staggering the pattern is a process known as _____.

_____ 13. A variety of wood block flooring is available and generally installed with a mastic or adhesive. Most wood block floors are prefinished. Which of the following statements concerning prefinished wood floors is *incorrect*?
   A. The finish is superior to that which is normally applied on the job site.
   B. Special care and accuracy must be maintained when making the installation.
   C. Time is saved since prefinished units can be butted directly against baseboards.
   D. Floors are ready for service immediately after the installation is completed.

_____ 14. Laminated strip flooring is a composite material whose core consists of medium- or high-density _____ sandwiched between layers of melamine. The top layer carries a wood-grained pattern.

15. Study the illustration of laminated strip flooring. What procedure is illustrated?

_____

_____

_____ 16. When installing engineered wood flooring, a couple of pieces of masking tape can be applied to hold pieces together until the _____ dries.

_____ 17. Conventional subflooring must be covered with an underlayment before the installation of resilient floor tile. Hardboard panels are often used for this purpose and should be conditioned to the space where they will be applied for at least _____ hours.
   A. 24
   B. 36
   C. 48
   D. 60

_____ 18. When plywood is used for underlayment, the grain of the panels should run at a right angle to the floor joists. Field spacing of nails for a 1/4″ thickness should not be greater than _____.

Name _____

_____ 19. The underlayment surface for resilient materials such as vinyl, rubber, and linoleum must be smooth. Over a period of time, even the slightest irregularities will show on the surface of the tile. This transfer from the base surface to the finish surface is commonly referred to as ____.

_____ 20. When stapling hardboard underlayment to the subfloor, nails or staples should be spaced 3/8" from the edge of the panels. The panels should be carefully fitted together with a(n) ____ space at each joint.

_____ 21. To check the right angles formed by the main centerlines of a room layout, you can use a framing square or set up a right triangle with a base of 4', an altitude of 3', and a hypotenuse of ____'.
A. 2
B. 4
C. 5
D. 7

_____ 22. To ensure a balanced layout, start laying resilient tile ____.
A. in a corner of the room
B. at the center of the room
C. at the center of the longest wall
D. at the center of the shortest wall

_____ 23. ____ tiles are similar to standard floor tile, except the adhesive is applied to the tile back at the factory.

_____ 24. Flexible vinyl flooring is fastened down only along edges and seams. It is generally available in a width of ____'.
A. 8
B. 9
C. 10
D. 12

_____ 25. Mosaic ceramic tile is any tile ____" square or smaller.
A. 1
B. 2
C. 4
D. 8

_____ 26. The material between individual tiles in a finished application is ____.
A. mortar
B. thin set
C. grout
D. mastic

Name _____ Date_____ Score _____

CHAPTER **20**
Stair Construction

Carefully study the chapter and then answer the following questions.

_____ 1. In the normal sequence of construction, main stairways are built or installed after interior wall surfaces are complete and finished flooring or _____ has been laid.

_____ 2. The main stairway in a residential dwelling should be _____ wide for easy movement of people and furniture.

3. There are several ways to describe or classify a stair. One way is to indicate whether or not it is enclosed by walls. With this in mind, identify the stair drawings below.

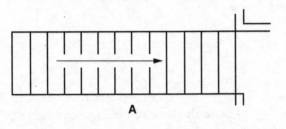

A

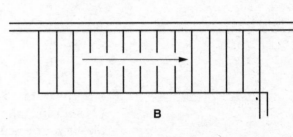

B

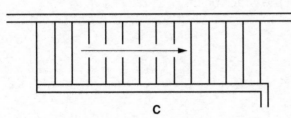

C

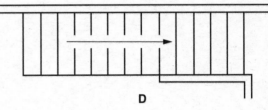

D

A. _____   C. _____

B. _____   D. _____

_____ 4. The size of the rough opening for a stairwell must be known or calculated during the rough framing of a structure. Members around the opening (trimmers and headers) should be doubled whenever they are longer than _____'.
A. 4
B. 6
C. 8
D. 10

5. Identify the basic parts and stair terms in the drawing below.

A. _____

B. _____

C. _____

D. _____

E. _____

F. _____

G. _____

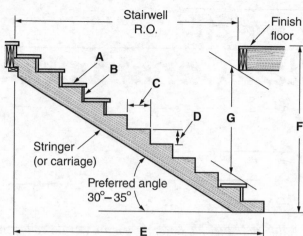

_____ 6. The minimum allowable headroom according to the IRC is ____.
   A. 5′-10″
   B. 6′-0″
   C. 6′-8″
   D. 7′-0″

_____ 7. The relationship or size ratio between risers and treads is very important in stair design. One commonly accepted rule states that the sum of two risers plus one tread should equal ____ inches.
   A. 18–20
   B. 20–22
   C. 24–25
   D. 25–26

_____ 8. If a given stair has a riser 6 3/4″ high, the correct width of the tread (less nosing) should be ____. Apply the rule referred to in question 7. Show your calculations in the space below.
   A. 10 1/2″–11 1/2″
   B. 11 1/2″–12 1/2″
   C. 13 1/2″–14″

Name _____

_____ 9. The IRC requires the minimum stair width of at least ____.
  A. 3'-6"
  B. 3'-0"
  C. 2'-10"
  D. 2'-6"

10. Except for very wide stairs, a handrail on one side or the other is satisfactory. The drawing below shows the position of a handrail attached to a wall. Provide the recommended heights for the labeled parts.

  A. _____

  B. _____

  C. _____

  D. _____

11. Calculate the number and size of risers and treads for a main stairway (straight run) for a residence. The vertical distance between the finished surfaces of the two floors is 8'-11" and the riser height must not be greater than 7 1/2". Use the riser-tread rule given in question 7. Show your calculations in the space below.

  No. of risers: _____

  Riser height: _____

  No. of treads: _____

  Tread width: _____

_____   12. Using the figures developed in question 11, determine the total run of the stairs. Show your calculations in the space below.

_____   13. Before making the actual layout of a stair stringer, the calculated riser height is laid out and checked with a straight strip called a(n) ____.

_____   14. When all of the risers and treads have been laid out on the stringer stock, an adjustment must be made for the thickness of the tread. This is best accomplished by ____.
   A. relocating each tread line downward by an amount equal to tread thickness
   B. extending the top riser and shortening the bottom riser by an amount equal to tread thickness
   C. shortening the bottom riser by an amount equal to the tread thickness

_____   15. As a general rule, when the basic width of a tread is increased, the nosing width is ____ (*increased, decreased*).

16. Identify the three basic types of riser design shown.

A ___   B ___   C ___

A. _____   C. _____

B. _____

_____   17. The simplest type of stringer is the ____ stringer.

Name _____

_____ 18. Open risers must include a structure that prevents a _____ from passing through the open space.
    A. person's foot
    B. child's head
    C. 6-inch cube
    D. 4-inch sphere

_____ 19. Winder stairs have a tapered tread in the section where the direction of the run is changed. The narrow end of the tread must be at least _____ inches wide.

20. Main stairs that are open on one or both sides usually have a decorative structure that supports the handrail. This assembly is generally called a balustrade. In the view below, identify the parts specified.

A. _____

B. _____

C. _____

D. _____

E. _____

_____ 21. Most building codes require that balusters be spaced no more than _____.

_____ 22. Which part of a balustrade must be securely attached to the starter step or anchored to the building frame?
    A. Open stringer.
    B. Baluster.
    C. Newel.
    D. Bracket.

_____ 23. Stair parts are ordered through a lumber dealer or a(n) _____ dealer.

_____ 24. The main purpose of _____ is to prevent anyone from slipping under the railings and falling to the floor below.

CHAPTER **21**

Doors and Interior Trim

Carefully study the chapter and then answer the following questions.

_____ 1. *True or False?* Prehung doors usually come with locks installed.

_____ 2. Standard interior doorjambs for walls framed with 2×4s are 1″ nominal thickness and ____″ wide.
   A. 4 5/8
   B. 5
   C. 5 1/4
   D. 5 3/8

3. Identify the parts of the door frame shown in the illustration below.

   A. _____

   B. _____

   C. _____

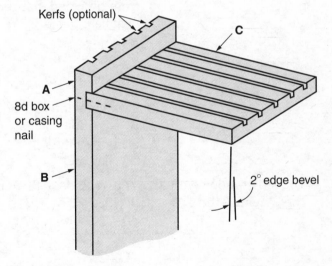

_____ 4. Side jambs are usually cut so they measure ____ below the head jamb. This length provides clearance under the door and any excess can be trimmed off.

_____ 5. When installing doorjambs, shims should be installed approximately ____ from the top.

_____ 6. The two general types of doors are panel and flush. The standard thickness of interior passage doors is normally ____.

7. The drawing below shows a section through a doorjamb. Provide the measurement specified and identify the parts.

A. _____

B. _____

C. _____

D. _____

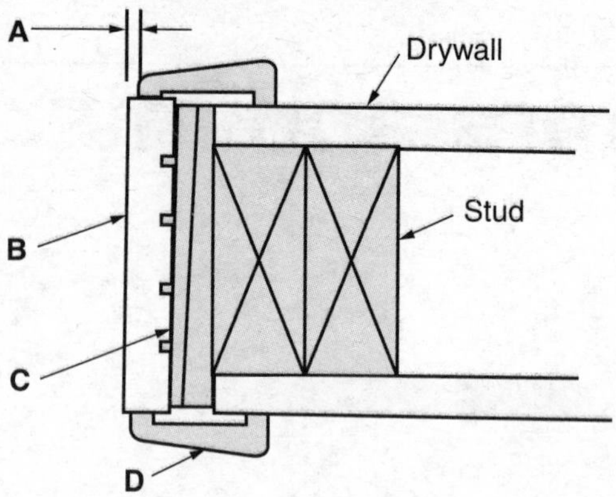

_____ 8. When using a miter joint between side and head casing for a door, which of the following installation procedures and requirements is *incorrect*?
   A. Hold the side pieces in place and mark the position of the miter joint at the top.
   B. Use a miter box or wood trimmer to cut the joint.
   C. Temporarily attach the side casings and then lay out and cut the head casing.
   D. Use 8d casing or finish nails to secure the casing to the jamb and space them about 10″ apart.

_____ 9. Panel doors consist of stiles, rails, and panels. The panels are made from plywood, hardboard, or thin solid stock. After the door is installed, the rails are in a _____ position.
   A. vertical
   B. horizontal
   C. diagonal

_____ 10. In the previous question, the stiles are in a _____ position.
   A. vertical
   B. horizontal
   C. diagonal

11. The illustration below shows two kinds of flush doors. One has a solid core, the other a hollow core. Identify the parts as specified.

A. _____

B. _____

C. _____

D. _____

E. _____

F. _____

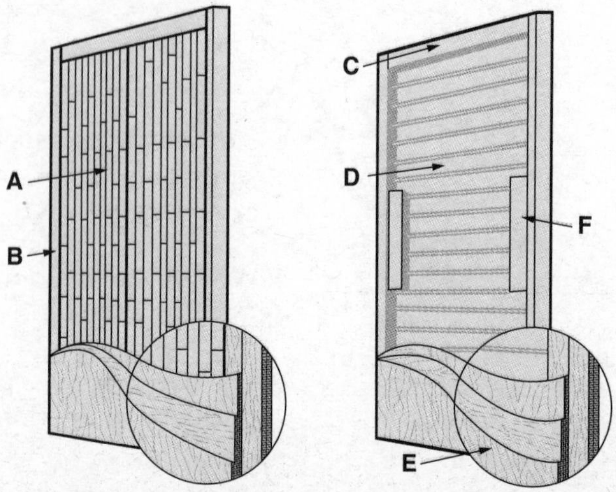

Name _____

12. Study the illustration of a door below and identify its parts.

   A. _____

   B. _____

   C. _____

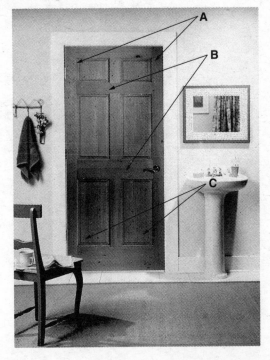

_____ 13. Doors are designed to fit standard-size openings and should not be cut to fit smaller openings. Cutouts for glass inserts in flush doors should not be more than _____ of the face area.
   A. 30%
   B. 40%
   C. 50%
   D. 60%

For questions 14–17, refer to the illustration below.

_____ 14. A power plane is being used to trim a door to size. The recommended clearances for an interior door are _____ on the lock side.
   A. 1/16″
   B. 3/32″
   C. 1/8″
   D. 3/16″
   E. 1/2″
   F. 5/8″

_____ 15. A power plane is being used to trim a door to size. The recommended clearances for an interior door are _____ on the hinge side.
   A. 1/16″
   B. 3/32″
   C. 1/8″
   D. 3/16″
   E. 1/2″
   F. 5/8″

_____ 16. A power plane is being used to trim a door to size. The recommended clearances for an interior door are _____ at the top.
   A. 1/16″
   B. 3/32″
   C. 1/8″
   D. 3/16″
   E. 1/2″
   F. 5/8″

_____ 17. A power plane is being used to trim a door to size. The recommended clearances for an interior door are _____ at the bottom.
   A. 1/16″
   B. 3/32″
   C. 1/8″
   D. 3/16″
   E. 1/2″
   F. 5/8″

Name _____

_____ 18. After the door has been trimmed to fit the opening, a bevel is planed on the lock side. This bevel should be about ____° (slightly greater for narrow doors and slightly less for wide doors).
A. 2
B. 2 1/2
C. 3
D. 3 1/2

_____ 19. Which statement best describes a common type of security hinge?
A. A protrusion on one leaf fits into a hole in the opposite leaf.
B. The hinge pin is designed so that it cannot be easily removed.
C. The hinge is completely concealed when the door is closed.
D. The hinge can only be removed with a special tool.

_____ 20. After the door and hinges have been mounted on the jamb and checked for proper clearance, the stops are installed. With the door in a closed position, the clearance of the stop on the hinged side should be ____".
A. 1/32
B. 1/16
C. 3/32
D. 1/8

21. Provide the correct name for the four types of passage door locks shown below.

A. _____

B. _____

C. _____

D. _____

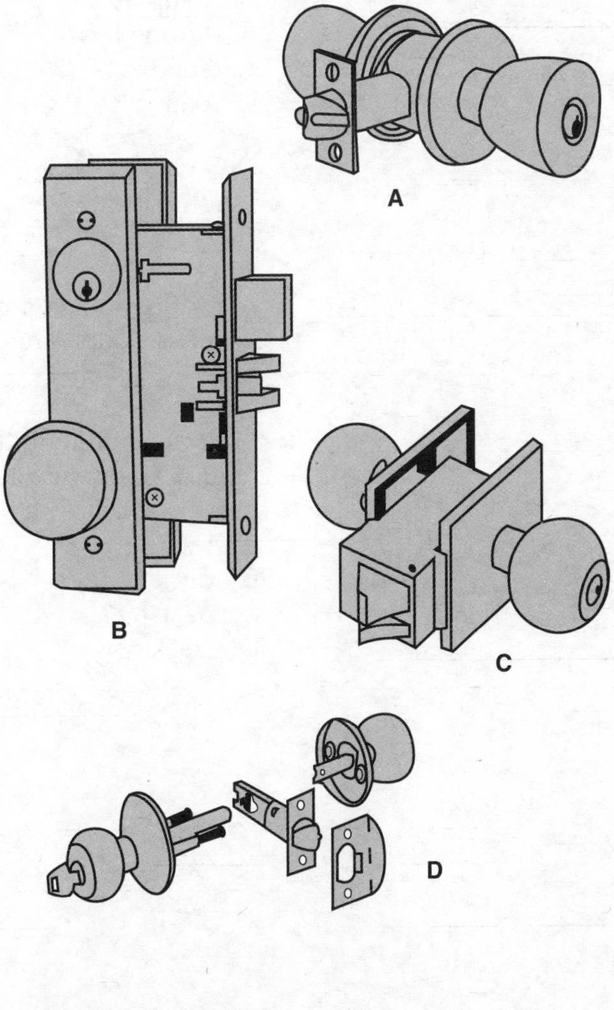

_____   22. To provide extra security, outside doors are often equipped with a special, additional lock. This lock, which may be keyed from both sides, is called a(n) ____.

23. When ordering locks, it may be necessary to specify the swing of the door. This is determined by facing the outside of the door. Determine the swing (the hand of the door) for the two drawings below. Give your answer in the accepted abbreviations of RH and LH.

A. _____

B. _____

A   B

_____   24. Passage door locks are normally installed in the door a vertical distance of ____ above the finished floor. This distance is measured to the centerline of the knob.

_____   25. The positions of centerlines for holes that must be bored to mount cylindrical locks are laid out with ____ furnished by the manufacturer.
A. templates
B. boring jigs
C. spacers
D. patterns

_____   26. One type of sliding door requires the installation of track during the rough framing. This type of door is commonly referred to as a(n) ____.

_____   27. A folding door commonly used on closets and wardrobes consists of a pair of doors hinged together. Doors of this type are commonly called ____ units.

_____   28. Multipanel folding doors are constructed from narrow panels that are hinged along the edges. As the door is opened, the panels fold together forming a ____.
A. bundle
B. deck
C. pack
D. stack

Name _____

29. Moldings are shaped strips of wood used to trim or connect various elements. Door casing, for example, connects the jamb to the wall surface. Shown below are typical moldings used for interior trim. Identify each item as specified.

A. _____

B. _____

C. _____

D. _____

E. _____

F. _____

G. _____

H. _____

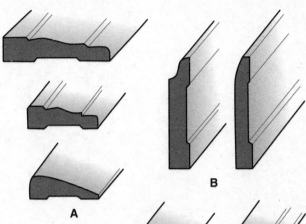
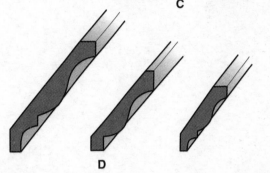
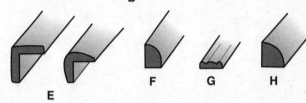

30. The drawing below shows a pair of standard, double-hung windows. Identify the interior trim members specified.

A. _____

B. _____

C. _____

D. _____

E. _____

F. _____

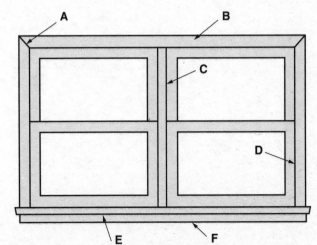

_____   31. When two pieces of crown molding must be joined end-to-end for a
long wall a _____ should be used.
A. butt joint
B. lap joint
C. scarf joint
D. dovetail

CHAPTER **22**

Cabinetry

Carefully study the chapter and then answer the following questions.

_____ 1. In building construction, the term *cabinetwork* refers to such items as kitchen cabinets, bathroom cabinets, and wardrobes. The term built-in means that the unit is _____ to the structure.

_____ 2. Cabinets may be built on the job by a carpenter, custom built in a cabinet shop, or _____ in a factory.

_____ 3. The positions of kitchen cabinets are usually shown on the _____.
   A. cabinet plans
   B. floor plan
   C. building elevations
   D. detail drawings

_____ 4. The typical height of a kitchen base cabinet before the countertop is installed is _____.
   A. 36″
   B. 36″-38″
   C. 32″
   D. 34 1/2″

5. The illustration below shows four standard types of cabinet door catches. Provide the correct name for each type.

   A. _____
   B. _____
   C. _____
   D. _____

6. The drawings below are section views of a kitchen cabinet, bathroom cabinet, and closet with sliding doors. Provide the recommended sizes as specified.

A. _____

B. _____

C. _____

D. _____

E. _____

F. _____

G. _____

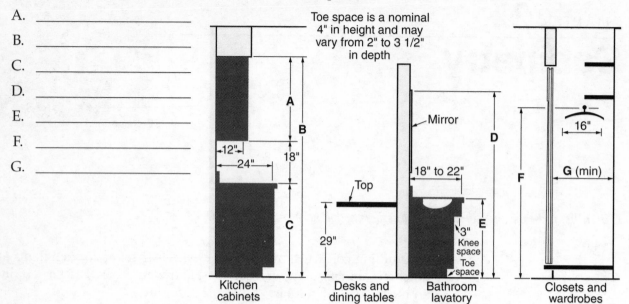

Toe space is a nominal 4" in height and may vary from 2" to 3 1/2" in depth

Kitchen cabinets

Desks and dining tables

Bathroom lavatory

Closets and wardrobes

_____ 7. Since floors are rarely level and walls seldom plumb, ____ and blocking are used on walls and floors so that cabinets do not become racked or twisted during installation.

_____ 8. Before installing cabinets, a level line at where the cabinet tops will be should be snapped on the wall where the cabinets will be located. This line should be ____.
A. measured up from the high point on the floor
B. measured down from the top line of the wall cabinets
C. the height of the cabinets from the low point on the floor
D. None of the above.

9. The drawing below illustrates steps to be taken for proper installation of cabinets so that they are not racked by uneven surfaces. Match the steps to the letters on the drawing. Place the numbers in the corresponding blanks.

1. Check the space for a tall unit by measuring up from the high point level.

2. Remove plaster at the high points.

3. Strike a level base line from the high point of the floor.

4. Tack on shims at low points or shim when attaching cabinets to the wall.

5. Mark the outlines of all cabinets on the wall to check the actual cabinet dimensions against your layout.

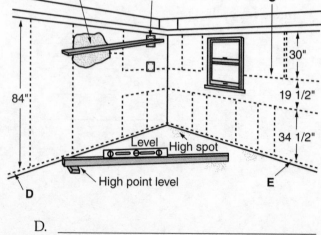

A. _____   D. _____

B. _____   E. _____

C. _____

Name _____

_____ 10. When the base cabinets are installed before the wall cabinets, the tops of the base cabinet can support a ____ to hold the wall cabinet in position for fastening.

11. Identify the three types of cabinet swinging doors shown below.

A. _____

B. _____

C. _____

_____ 12. *True or False?* High-quality cabinets have doors made entirely of solid wood.

_____ 13. *True or False?* Euro-style hinges are easy to install.

_____ 14. The hinge shown in the illustration below is designed for use on a ____ door.
A. flush
B. lipped
C. overlay
D. All of the above.

_____ 15. Drawer pulls usually look best when they are installed ____.
A. slightly above the center of the drawer front
B. centered between the top and bottom of the drawer front
C. slightly below the center of the drawer front

_____ 16. What type of hinge is shown in the photo below?
A. Knife.
B. Butt.
C. Euro-style.
D. Cup.

_____   17. When wood or metal pins are used to support adjustable shelves, it is recommended that the holes for the pins be drilled ____ (*before, after*) the basic cabinet is assembled.

_____   18. Standard shelving that is 3/4" thick should be carried on supports that are spaced not more than ____" apart.
A. 32
B. 36
C. 40
D. 42

_____   19. When standard butt hinges are used to install a cabinet door, they are usually mounted in a cutout called a ____.
A. rabbet
B. recess
C. mortise
D. router

_____   20. Which of the following is *not* used for countertops?
A. Glass.
B. Soapstone.
C. Synthetic resin.
D. All of the above are used.

_____   21. What is *not* a characteristic of granite countertops?
A. Scratch resistant.
B. Resists staining by colored liquids.
C. Not affected by heat of cooking.
D. Available in a range of colors and patterns.

_____   22. Which of the following is a component of engineered quartz countertops?
A. Synthetic resin.
B. Granite chips.
C. Glass.
D. All of the above.

_____   23. Solid-surface countertops should be supported every ____ inches.

_____   24. Plastic laminate commonly used for the surface of cabinet counters and tops is ____" thick.
A. 1/32
B. 1/16
C. 3/32
D. 1/8

_____   25. On-the-job installations of plastic laminates are normally made with an adhesive called ____.
A. urea resin glue
B. contact bond cement
C. polyvinyl glue
D. casein waterproof glue

Name _____

_____ 26. Architectural Woodworking Institute standards specify that a backing
sheet of plastic laminate be used on any unsupported area of
counters or tops that exceed _____ sq. ft.
A. 2
B. 3
C. 4
D. 6

27. Describe *build-up strips* and explain how they are used.

_____

_____

CHAPTER **23**

Painting, Finishing, and Decorating

Carefully study the chapter and then answer the following questions.

_____ 1. Anyone who paints, renovates, or repairs a structure built before 1978 must be certified by the _____ or an authorized state agency.

_____ 2. VOC stands for_____.
   A. Volume Over Content
   B. Very Organic Compounds
   C. Volatile Organized Content
   D. Volatile Organic Compounds

_____ 3. *True or False?* Paint and clear finishes are the only materials that contain VOCs.

_____ 4. _____ should be used to fill cracks and small openings before painting, but it does not perform well where it may be stretched by expansion and contraction.

5. Identify the specified parts of the paintbrush illustrated below.

   A. _____
   B. _____
   C. _____
   D. _____
   E. _____

_____ 6. The purpose of the plug in a paintbrush is to _____.
   A. save on the number of bristles in the brush
   B. create a reservoir that will hold a supply of paint
   C. provide a base for attaching bristles

_____   7. The purpose of primer is to ____.
A. improve adhesion of the paint to the base material
B. improve coverage of the finish coat
C. provide an inexpensive base coat
D. help the wood absorb the solvents from the top coat

_____   8. What information may be found on a paint can label?
A. Coverage rate.
B. Drying time.
C. VOC level.
D. All of the above.

_____   9. The ____ type of paint sprayer uses the same principle of operation as an atomizer.

_____   10. Sandpaper sometimes used between coats of varnish or clear sealer to ____.
A. smooth previously coated surface
B. provide a "tooth" for better adhesion
C. reduce raising of the wood grain

_____   11. ____ is the lightness or darkness of a color.

_____   12. The illustration below shows a grid of expanded metal placed in a paint container. Its purpose is to ____.
A. flatten the nap of a roller so paint will not spatter
B. provide a place to strike off excess paint from a brush or roller
C. place a pattern on the roller that will be transferred to the surface being painted

_____   13. *True or False?* Alkyds and water-based paints require more brushing than oil paints.

Name _____

_____ 14. A dent in a surface being stained and varnished should be removed before beginning to stain. A small dent may be raised by applying a small amount of ____.
A. water
B. alcohol
C. shellac

_____ 15. *True or False?* Extremely weathered exterior wood surfaces must be sanded to "bright" wood before being repainted.

_____ 16. A painted surface develops blisters and begins to peel within a year after application. The probable cause is ____.
A. inferior paint that lacks elasticity
B. water under the coating is pushing the film away from the wood surface
C. the use of too little oil in the paint
D. the formation of mildew

_____ 17. The illustration below shows paint failure called checking. This type of paint failure is caused by ____.
A. poor-quality paint that lacks elasticity
B. paint that was applied too thick
C. ultraviolet rays causing oils to break down
D. hard finishing coat applied over a softer primer

18. Using the following information, figure out how much paint is required to cover the walls of a 12′ × 14′ room with 8′ ceilings. The windows and doors take up 48 sq. ft. and the paint coverage is 400 sq. ft. per gallon. Two coats will be applied. Show your calculations in the space below.

CHAPTER **24**

Chimneys and Fireplaces

Carefully study the chapter and then answer the following questions.

_____ 1. Masonry chimneys are usually freestanding, which means they are structurally separate from the building frame. Footings should extend below the frost line and project at least ____" beyond the sides.
A. 2
B. 3
C. 4
D. 6
E. 8
F. 10

_____ 2. The walls of a chimney with a clay flue lining should be at least ____" thick.
A. 2
B. 3
C. 4
D. 6
E. 8
F. 10

3. Building codes require that chimneys be constructed high enough to avoid downdrafts caused by wind turbulence. In the illustration below, provide the minimum recommended dimensions for each chimney.

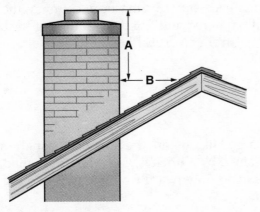

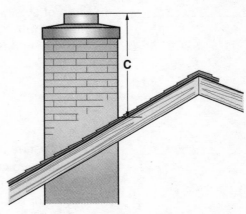

A. _____

B. _____

C. _____

Use the drawing below to answer Questions 4–5.

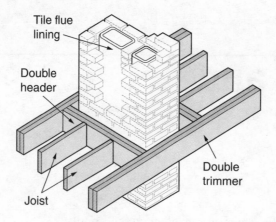

_____ 4. The drawing above shows a cross section of a chimney surrounded by typical framing. The framing should clear the masonry by a minimum distance of ____.

_____ 5. When two flues adjoin each other, as shown, the joints of the flue lining should be vertically offset a minimum distance of ____.

_____ 6. Flue lining for furnaces, gas fireplaces, gas stoves, pellet stoves, and water heaters can be made of any material that is approved by ____.
A. International Building Code
B. Underwriters Laboratories, Inc.
C. International Residential Code
D. National Fire Protection Association

_____ 7. The area of passage (inside cross section) of a 12 × 16 modular flue liner is listed as 120 square inches. If a round liner is to be substituted for the 12 × 16 modular liner, what size should be selected?
A. 10" dia.
B. 12" dia.
C. 15" dia.
D. 18" dia.

_____ 8. When offsets or bends are necessary in a masonry chimney, the flue lining should be carefully mitered and fitted. The angle of the bend should never exceed ____° from a vertical line or plane.
A. 30
B. 45
C. 55
D. 60

_____ 9. Chimneys are often enlarged just before they project through the roof. This enlargement is obtained by a brick laying operation called corbeling, and should extend downward from the roof framing by a distance of not less than ____".
A. 2
B. 4
C. 6
D. 8

Name _____

10. The drawing below shows a section view of a standard masonry fireplace. Identify the parts as specified.

A. _____

B. _____

C. _____

D. _____

E. _____

F. _____

G. _____

_____ 11. The _____ is the part of a fireplace in which the ashes are collected and stored for later removal.

_____ 12. The metal structural member ordinarily placed across the top of the fireplace opening to support further masonry construction is called a(n) _____.
A. angle iron
B. lintel
C. head jamb
D. header

_____ 13. The side and back walls of the fireplace (where the fire is located) must be lined with firebrick up to the level of the _____. The firebrick must be set in a special clay mortar.

_____ 14. The sidewalls of a standard fireplace are usually constructed at an angle, mainly for the purpose of reflecting heat into the room. This angle is normally laid out at about _____" per foot.
A. 3
B. 4
C. 5
D. 6

_____ 15. The passageway between the combustion chamber (main fireplace) and the smoke chamber is called the throat. The throat _____ controls the efficiency of a fireplace.
A. length
B. size
C. width
D. depth

_____ 16. When installing a damper unit, a clearance should be provided at each end to allow for ____.
 A. soot buildup
 B. free operation
 C. insulation
 D. expansion

_____ 17. Most surfaces of the smoke chamber are plastered with a coat of cement mortar about ____″ thick.
 A. 3/8
 B. 1/2
 C. 5/8
 D. 3/4

_____ 18. One recommended method of calculating the cross-sectional area for a fireplace flue is to allow ____ sq. in. for every 1 sq. ft. of the fireplace opening.
 A. 7
 B. 9
 C. 11
 D. 13

_____ 19. A somewhat ____ (_larger, smaller_) flue may be required in chimneys lower than 20′.

_____ 20. The purpose of the smoke shelf is to redirect the downdraft of the chimney. It should be made equal to or longer than the damper unit and no less than ____″ in depth.
 A. 4
 B. 6
 C. 8
 D. 10

_____ 21. The use of a metal built-in circulator increases the heating efficiency of a fireplace. Standard models are designed in about the same way as a regular masonry fireplace. They include a combustion chamber, throat, damper, smoke shelf, and ____.

_____ 22. The housing of a ____ fireplace can rest directly on a wood floor.

_____ 23. The metal chimney pipe system for a factory-built fireplace should be supported ____ so the entire weight of the chimney does not rest on the fireplace.
 A. at the roof
 B. by braces the manufacturer supplies
 C. every 6 feet
 D. All of the above.

_____ 24. A box-like structure built either inside or outside of the regular building framing to hold the fireplace or chimney system is called a(n) ____.

Name _____ Date _____ Score _____

CHAPTER **25**

Post-and-Beam Construction

Carefully study the chapter and then answer the following questions.

1. Identify the framing members of post-and-beam construction specified in the drawing below.

   A. _____

   B. _____

   C. _____

   D. _____

   E. _____

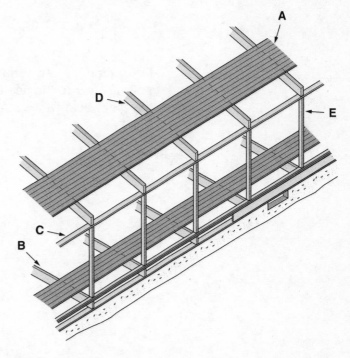

_____ 2. An important advantage of post-and-beam construction over standard wood framing or frames consisting of metal beams is its high resistance to ____.

_____ 3. Beams are often joined by butting them together and locating the joint over a post. The bearing surface of the post should be increased by attaching ____ or using a heavy steel plate.
   A. angle irons
   B. beam hangers
   C. bearing blocks
   D. metal straps

_____   4. Determine the 1/d ratio for a 4 × 4 solid post that is 10′ long. Round your answer to the nearest whole number. Show your calculations in the space below.

_____   5. Determine the 1/d ratio for a 6 × 6 solid post that is 14′ long. Round your answer to the nearest whole number. Show your calculations in the space below.

6. Beams may consist of solid wood or may be built up in various ways. Identify the types of beams shown in the cross-sectional views below.

A. _____

B. _____

C. _____

D. _____

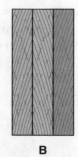

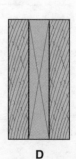

A   **B**   **C**   **D**

Name _____

_____ 7. Posts are usually evenly spaced along the length of the structure. To take full advantage of modular materials, the spacing should be based on standard increments of ____", 24", and 48".
A. 16
B. 18
C. 20
D. 30

8. Typical sill construction for a post-and-beam frame is shown in the illustration below. Identify the parts specified.

A. _____

B. _____

C. _____

D. _____

E. _____

F. _____

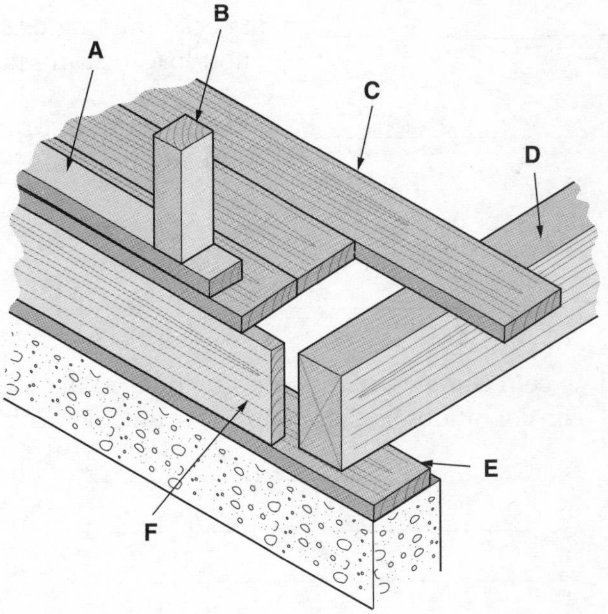

_____ 9. There are two basic types of roof beams. A transverse beam runs in the same direction as a common rafter. The other type runs parallel to the supporting side walls and is called a(n) ____ beam.

10. The illustrations below show two methods of building a plank-and-beam roof. Name each type.

A. _____

B. _____

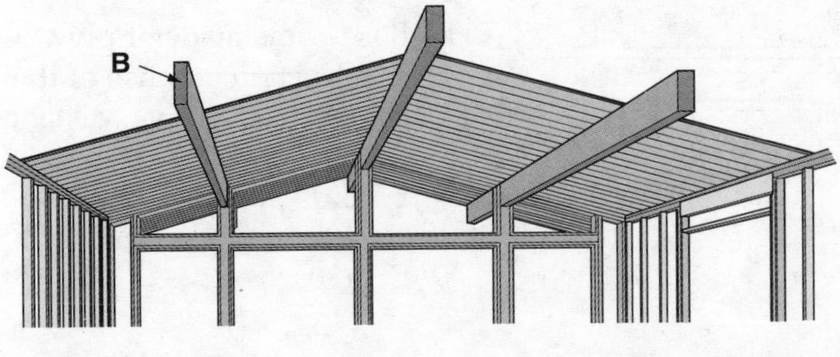

_____   11. A post-and-beam frame consists of a limited number of joints. Metal connectors are often used to reinforce these joints. To increase the holding power of the metal connectors, they should be attached with bolts or ____.

_____   12. A partition that runs parallel to a transverse beam has a ____ top plate.
A. staggered
B. horizontal
C. supporting
D. sloping

_____   13. Special framing or support ____ (*is, is not*) required when non-load-bearing partitions run parallel to the floor planks.

_____   14. Planks for floor or roof decks usually have a tongue-and-groove edge and are sometimes end matched. Standard thicknesses vary from ____.
A. 1 1/2″ to 3″
B. 2″ to 3″
C. 2″ to 4″
D. 2″ to 4 1/2″

15. In cold climates, plank roof structures located directly over heated areas require the same considerations as those applied to outside walls. Identify the items specified in the drawing below.

A. _____
B. _____
C. _____
D. _____

Plank roof deck

Blocking

Exposed beam

4 x 4 post directly under beam

_____   16. ____ can be designed to carry structural loads over wide spans.

_____   17. Box beams made of plywood webs are often used in post-and-beam construction because of their high strength-to-weight ratio. They can be designed to span distances up to ____′.
A. 60
B. 80
C. 100
D. 120

Name _____

_____ 18. Laminated wood beams and arches are usually made from layers of _____ (*hardwood, softwood*) glued together with waterproof adhesives.

19. In residential construction, laminated beams are generally straight or tapered. In institutional or commercial buildings, however, they are often formed into curves, arches, and other shapes. Provide the correct name for the standard forms shown below.

A. _____

B. _____

C. _____

D. _____

E. _____

F. _____

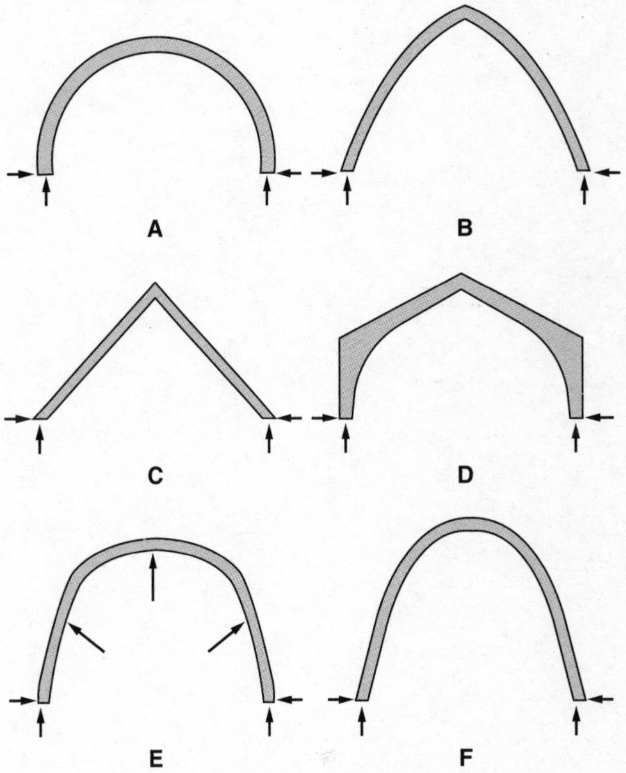

_____ 20. When fabricating long lengths of laminated arches and beams, it is usually necessary to join the ends of pieces that make up a given layer. A special finger joint is commonly used. These joints should be staggered at least _____" in adjacent layers.
A. 16
B. 24
C. 32
D. 36

Name _____ Date _____ Score _____

CHAPTER **26**

Systems-Built Housing

Carefully study the chapter and then answer the following questions.

_____ 1. The terms systems built and factory built refer to the cutting and assembly of parts, subassemblies, and sections in plants and factories. These units are then transported to the ____ for final assembly and erection.

_____ 2. In systems- or factory-built housing, three-dimensional units that are fully assembled before they leave the manufacturing plant are called ____.

_____ 3. In roof truss construction, ____ saws can be used to cut members to length.
   A. movable table
   B. radial arm
   C. high-production
   D. swinging cutoff

4. Basically, there are five types of prefabricated buildings. Match each of the brief descriptions with its corresponding type.

_____ Modular

_____ Precut

_____ Manufactured Home

_____ Log Home

_____ Panelized

A. Flat sections of the structure are fabricated on assembly lines.

B. All lumber is cut, shaped, and labeled.

C. Entire units are assembled and finished on both the inside and outside.

D. The floor frame is attached to a steel chassis.

E. Logs are precut and stacked.

5. The illustration below shows a particular system that provides ample space for installation of plumbing lines. What is the name of this system?

_____

_____   6. The various members of a roof truss are usually assembled with ____ connectors.
   A. split-ring
   B. gang-nail
   C. chord-bolt
   D. T-bolt

_____   7. In ____ prefabrication, flat sections of the structure are built on assembly lines.

_____   8. In manufacturing plants, wall panel frames are assembled by placing the studs, headers, and plates in positioning ____ and then fastening them together with pneumatic nailers.

_____   9. Closed panels may have voids or channels through which electrical wiring or ____ are run.

_____   10. In the view below, a gang of ____ is being used to secure plywood sheathing to an outside wall frame.
   A. power nailers
   B. motorized drills
   C. electric staplers
   D. glue applicators

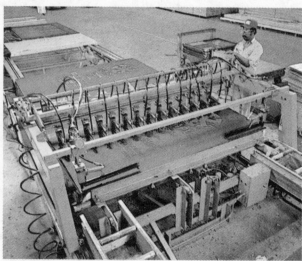

Name _____

_____ 11. In panelized prefabrication, a minimum amount of onsite labor is required to complete the structure. Which of the following statements is *incorrect* concerning this type of construction?
    A. Most of the wall, ceiling, and floor surfaces are finished in the plant.
    B. Kitchen cabinets can be installed before the section leaves the plant.
    C. Plumbing fixtures are seldom installed until erection is completed on the building site.
    D. Transportation limits the maximum width of any given section.

_____ 12. A section of modular construction that includes a concentration of heating and plumbing facilities is often included in a prefabrication system that consists mainly of panels. The section is generally referred to as a(n) _____.

_____ 13. Manufactured homes are over 320 square feet and have a _____ permanently attached.
    A. manufacturer's name plate
    B. wheeled chassis
    C. trailer hitch
    D. leveling mechanism

_____ 14. What building product is being used to assemble the walls of the house in this photo?

_____ 15. *True or False?* Modular homes are limited to one or two stories.

Name _____   Date_____   Score _____

CHAPTER 27

Green Building and Certification Programs

Carefully study the chapter and then answer the following questions.

_____ 1. Which of the following statements about green homes is true?
  A. Green homes are as comfortable as traditional homes.
  B. Green homes are built to be durable and require as little maintenance as possible.
  C. A green home is designed to conserve energy and water.
  D. All of the above.

_____ 2. *True or False?* Green homes are more expensive to construct and operate, but have less impact on the environment.

_____ 3. What is the most important aspect of a green home?
  A. Having all parts work together.
  B. Saving energy.
  C. Using appliances that conserve water.
  D. Both B and C.

_____ 4. Which of the following should be members of the green construction team?
  A. Designer/architect.
  B. Owner.
  C. Subcontractors.
  D. All of the above.

_____ 5. As shown in the figure below, the _____ shades the windows, reducing solar heat from entering the building.

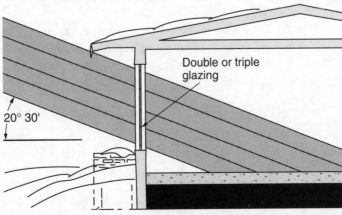

Double or triple glazing

20° 30'

Sun angle—12:00 noon December 22
45° north latitude

_____   6. For a house to receive the maximum amount of winter heat from the sun, which side should have the most windows?
 A. North.
 B. East.
 C. South.
 D. West.

_____   7. Locating the structure to receive maximum advantage of solar heating is called determining its _____.
 A. balancing
 B. orientation
 C. sizing
 D. face

_____   8. How does the location of the mill or factory where a product is produced affect the green characteristics of a house?
 A. Materials that are transported shorter distances, use less transport fuel.
 B. Some localities have stronger regulations to protect the environment than others.
 C. Wood products that come from areas with large forest reserves do less to impact the environment.
 D. Shorter transportation distances help keep the costs of construction down.

_____   9. What are the advantages of engineered lumber for green home construction?
 A. Longer, straighter pieces are available.
 B. Less waste.
 C. More dimensionally stable.
 D. All of the above.

_____   10. Which of the following is *not* a significant cause of poor indoor air quality?
 A. High VOC paint.
 B. Poor flashing around windows and doors.
 C. Poorly designed heating and air conditioning system.
 D. Carpeting.

_____   11. Which of the following is a consideration in choosing green building products?
 A. Can it be recycled at the end of its useful life?
 B. Is it made from recycled materials?
 C. Is it made from sustainable materials?
 D. All of the above.

_____   12. The greatest percentage of energy consumption in a modern, well-constructed home is for _____.

Name _____

_____ 13. Devices with the Energy Star label consume _____ less energy than standard models.
A. 10-15%
B. 20-30%
C. 30-50%
D. at least 50%

_____ 14. *True or False?* A hole found in a outside pipe needs to be fixed by the plumbing crew and does not need to be communicated to the rest of the team.

_____ 15. Using WaterSense efficient shower heads can save an average family about _____ gallons of water per year.
A. 1,400
B. 2,000
C. 2,900
D. 4,100

_____ 16. What measures can be taken to reduce unacceptable levels of radon in a house?
A. Seal foundation cracks.
B. Increase ventilation in the area of build-up.
C. Place a vapor barrier between the foundation and the ground.
D. All of the above.

17. What is necessary for mold to grow?

_____

_____ 18. Which of these is *not* an NGBS level?
A. Ruby.
B. Emerald.
C. Silver.
D. Gold.

Name _____ Date _____ Score _____

Remodeling, Renovating, and Repairing

Carefully study the chapter and then answer the following questions.

_____ 1. When remodeling requires changes in the structure of a building, the carpenter should attempt to _____ the type of framing originally used.

2. The following list details some of the jobs that must be performed to renovate the exterior of an old house. Place them in a logical order.

_____ 1.   A.   Repair or replace windows.

_____ 2.   B.   Demolish construction that is to be changed and dispose of the debris.

_____ 3.   C.   Repair or replace the roof.

_____ 4.   D.   Perform all structural work, proceeding from the bottom up.

_____ 5.   E.   Stain or prime wood siding.

_____ 6.   F.   Paint.

_____ 7.   G.   Caulk, glaze, and putty.

_____ 8.   H.   Regrade the site and provide drainage away from the house.

3. Identify the parts of a balloon frame, as shown in the illustration below.

A. _____

B. _____

C. _____

D. _____

E. _____

F. _____

Exterior wall   **Standard sill**

4. Describe what is being done in the illustration below.

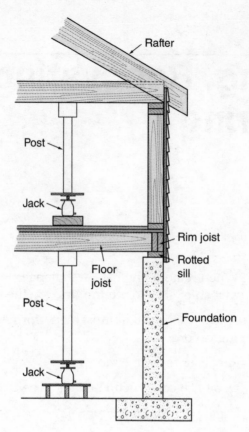

Labels on illustration: Rafter, Post, Jack, Floor joist, Post, Jack, Rim joist, Rotted sill, Foundation

_____

_____

_____ 5. When remodeling work includes excavating, the plans should include the location of _____ utility services.

_____ 6. For the rapid removal of lath and drywall, many carpenters use a(n) _____.
   A. reciprocating saw
   B. garden spade
   C. rip chisel
   D. wrecking bar

Name _____

7. Identify the framing members in the following illustration.

A. _____

B. _____

C. _____

D. _____

E. _____

F. _____

G. _____

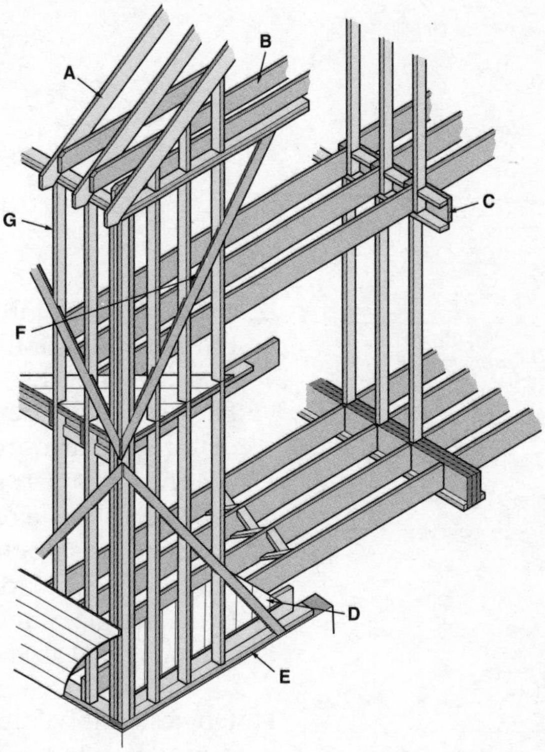

_____ 8. Shoring must be erected before removing a bearing wall. Which one of the following statements is *not* an indication that the wall is load bearing?
   A. Overhead joists are spliced over the wall.
   B. The wall runs parallel to floor and ceiling joists.
   C. The wall runs at right angles to overhead joists.
   D. The wall runs lengthwise down the middle of the building.

_____ 9. Headers in bearing walls must be sized to carry the load of the structure above. What is the best way to determine the appropriate size for a header?
   A. Look it up in the IRC or local building code.
   B. The depth should be 1 inch for every foot of opening.
   C. Use 2×8s for openings under 6 feet and 2×10s for wider openings.
   D. Use the header table on a framing square.

_____ 10. The short studs that support the header are called ____.

_____   11. The drawing below shows one method of supporting ceiling joists with a concealed header. The metal devices used for this construction are called ____.

_____   12. *True or False?* Jamb clips should be fastened to the doorjamb and then the assembly should be installed in the rough opening.

_____   13. How can badly damaged wood shingles can be replaced without disturbing surrounding shingles?
   A. Slide a flat bar under the shingles to carefully remove the nails.
   B. Use a shingle remover.
   C. Split the shingle where nails are located, then cut the nails off with a hacksaw blade.
   D. A shingle cannot be removed without removing or lifting the shingle(s) above it.

_____   14. Homes built before 1978 may contain ____ which can only be disposed of by a person who has been certified for such work.

Name _____ **Date** _____ **Score** _____

CHAPTER **29**

Building Decks and Porches

Carefully study the chapter and then answer the following questions.

_____ 1. A(n) _____ deck has one of its sides secured to a building.

_____ 2. *True or False?* A ground-level deck does not require a railing.

_____ 3. Resistance of building materials to _____ is a major concern when selecting deck materials.

4. Why is CCA-treated lumber no longer produced for use as a wood preservative in residential construction?

_____

5. When sawn lumber is used for decking, the bark side of the board should be the top surface. On the drawing below, which side of the board is the bark side—A or B?

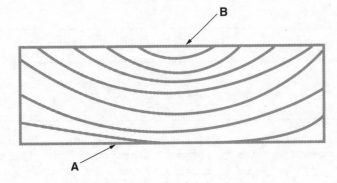

6. What are *glulams* and how are they manufactured?

_____

_____

_____

_____ 7. _____ are boards or planks manufactured from recycled plastics and wood fibers.

8. What measures are taken to prevent formation of rust and corrosion on metal fasteners used on decks?

_____

9. The method shown below is used to check whether deck corners are square. Provide the dimensions, in feet, indicated by the letters.

    A. _____

    B. _____

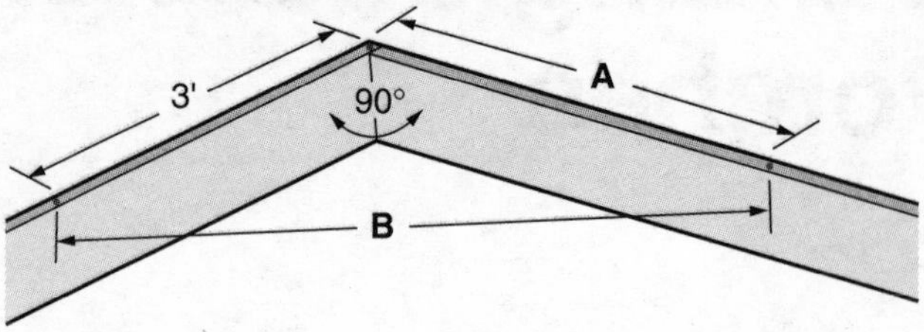

10. What are the four important steps in preparing a site for a deck?

_____

_____

_____

_____ 11. *True or False?* A deck is not a structural part of the house, so it is not necessary for deck piers to extend below the frost line.

_____ 12. A(n) ____, attached to a house, acts like a beam to support one side of a deck.

13. What are the three basic components of stairs.

_____

_____

_____

14. What are *stringers*?

_____

_____

_____

_____ 15. To allow for proper drainage, porch floors should slope away from the house ____" in 8 feet?
    A. 1/8
    B. 1/4
    C. 1/2
    D. 1

_____ 16. The three ____ styles are open, semi-enclosed, and closed.

CHAPTER **30**
Electrical Wiring

Carefully study the chapter and then answer the following questions.

_____ 1. Mechanical systems should be installed ____ (*before, after*) the building is insulated.

_____ 2. *True or False?* The National Electrical Code has the force of law in all municipalities of the United States.

3. What can be determined by inserting a neon tester's terminals inserted into the slots of a receptacle, as shown below?

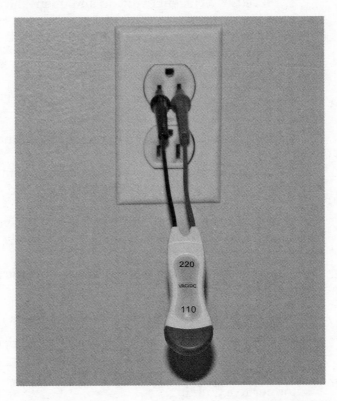

_____

_____

4. Identify the tool in the illustration below and explain how it is used when installing electrical service in a building.

_____

_____

_____   5. Which statement is true of ground-fault circuit interrupters (GFCIs)?
    A. It compares the current in the hot conductor with the voltage at the source.
    B. It compares the voltage in the hot conductor with the voltage in the neutral conductor.
    C. It compares the current in the hot conductor with the current in the neutral conductor.
    D. It senses a current overload and opens the circuit.

_____   6. What device detects erratic forms of electrical current?
    A. Arc-fault current interrupter.
    B. Ground-fault current interrupter.
    C. Circuit breaker.
    D. Ampredyne.

_____   7. Conductors are wires in an electrical circuit that carry _____.

_____   8. Electrical devices that protect conductors from overloads by shutting off electrical power to the circuit are _____ and circuit breakers.

_____   9. *True or False?* A step-up transformer receives voltage at a higher level and changes it to a lower voltage.

_____   10. Which ladder is *not* safe to use around overhead electrical wires?
    A. Wood.
    B. Aluminum.
    C. Fiberglass.
    D. Only wood ladders are safe around electricity.

_____   11. The _____ is the conductor that brings electrical power from a transformer to a building.

Name _____

12. Identify the electrical device represented by each of the electrical symbols below.

 A. _____

 B. _____

 C. _____

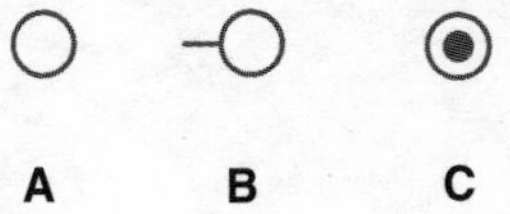

13. Identify the specified items shown in the illustration below.

 A. _____

 B. _____

 C. _____

 D. _____

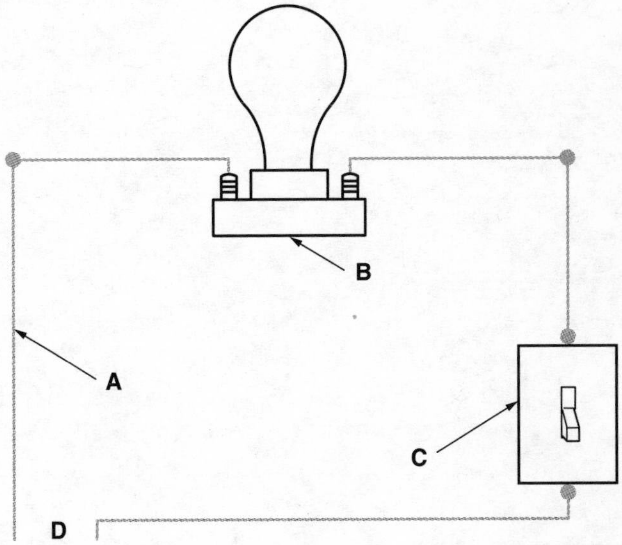

_____   14. Switches are placed only in the _____ of a circuit.
   A. hot conductor
   B. neutral or white conductor
   C. bare ground conductor

_____   15. *True or False?* If a test shows that there is no electrical current at the slots of a receptacle, it always means that the receptacle is faulty.

CHAPTER **31**

Plumbing Systems

Carefully study the chapter and then answer the following questions.

1. If framing members are notched to accommodate a plumbing pipe, what must be done after the pipe is installed?

   _____

   _____

   _____

2. Name the two plumbing subsystems.

   _____

   _____

_____   3. ____ is the part of the drainage piping that allows air to circulate in the pipes.

_____   4. Which of the following materials is *not* used for supply pipes and fittings?
   A. Copper.
   B. Galvanized steel.
   C. Malleable iron.
   D. Plastic.

_____   5. The two types of plastic used in plumbing hot-water supply piping are CPVC and ____ (use abbreviation).

6. What is the device shown in the photo below, and what is it used for?

_____

_____   7. Where is a wax ring used in plumbing a house?
  A. Between a faucet and the lavatory deck.
  B. Between a water closet and the closet flange.
  C. Between a sink drain and the sink basin.
  D. In sealing compression fittings.

8. Identify the parts of this compression valve.

  A. _____

  B. _____

  C. _____

  D. _____

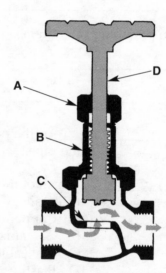

_____   9. Water-using devices in a house are called ____.

_____   10. *True or False?* Architectural drawings always include pipe drawings
    for plumbers to use as a guide to plumbing installations.

_____   11. *True or False?* Compression or flare fittings do not require soldering.

_____   12. To avoid crimping copper tubing while making a bend, a(n) ____
    should be used.

_____   13. Wells today are usually drilled or ____.

Name _____

14. What is the purpose of a pressure tank in a water supply system that is fed by a well?

_____

_____

_____ 15. Which step comes first when sweat soldering copper pipe?
    A. Heat the pipe just enough to soften the solder.
    B. Clean the pipe and fitting with steel wool or a cleaning brush.
    C. Apply paste flux to the pipe and fitting.
    D. Slip the fitting onto the pipe.

16. What should be applied to the threads of galvanized pipe before connecting a fitting?

_____

Name _____   Date _____   Score _____

CHAPTER **32**

Heating, Ventilation, and Air Conditioning

Carefully study the chapter and then answer the following questions.

_____ 1. *True or False?* The Second Law of Thermodynamics is that heat always moves from hot to cold.

_____ 2. The standard used for rating the efficiency of furnaces and boilers is AFUE. What does AFUE stand for?
   A. Annual Fuel Universal Energy.
   B. Actual Fuel Utilization Energy.
   C. Annual Fuel Utilization Efficiency.
   D. Average Fuel Utilization Engineering.

3. If a furnace consumes 1000 Btu of energy and is 85% efficient, how many Btu are wasted?

_____

4. Identify the specified parts of the gas furnace pictured below.

   A. _____
   B. _____
   C. _____
   D. _____
   E. _____

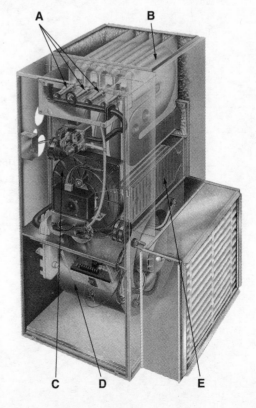

_____ 5. Which of the following items is *not* a part of a forced-air perimeter heating system?
   A. Boiler.
   B. Heat exchanger.
   C. Plenum.
   D. Blower.
   E. Burner.
   F. Ductwork.

_____ 6. In a forced-air heating system, a(n) _____ is a system of ducts that brings cold air back to the furnace from various rooms of the building.

7. Name two types of materials that are satisfactory for ductwork in a warm-air perimeter system.

_____

8. Identify the specified parts of the system below.

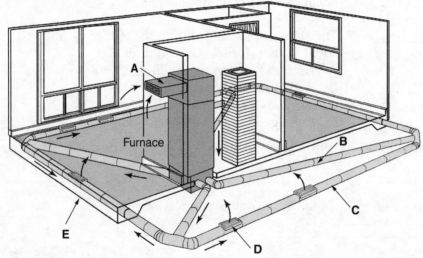

Furnace

A. _____   D. _____

B. _____   E. _____

C. _____

_____ 9. A manifold is used only in a hydronic heating system if _____.
   A. there is an adjusting valve
   B. the system is a two-pipe system
   C. the system is a one-pipe system
   D. the building has zone heating

Name _____

10. Identify the parts of the basic hydronic system shown in the drawing below.

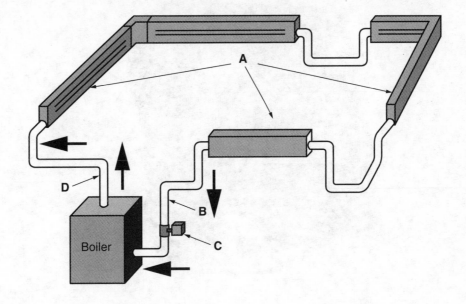

A. _____   C. _____

B. _____   D. _____

_____   11. A(n) _____ is located in the water supply to a hydronic system to limit the incoming water pressure.

_____   12. What type of heating system uses hot water pumped through tubing embedded in a concrete slab or beneath subflooring?
A. Hydronic radiant heating system.
B. Hydronic two-pipe system.
C. Floor convection system.
D. Quick recovery system.

_____   13. In an air-cooling system, there are two coils through which refrigerant is circulated. One coil, the cooling coil, absorbs heat from the warm air passing over it. The heated refrigerant is then pumped through another coil, the _____, where the refrigerant passes its collected heat to the atmosphere.

14. Name the device represented by the drawing below.

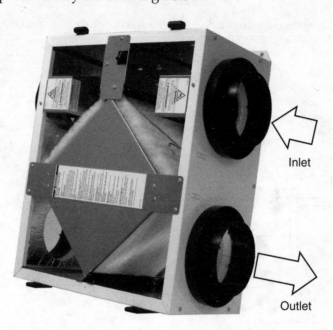

Inlet

Outlet

_____

_____ 15. Testing has shown that indoor air should be exchanged with fresh
outdoor air ____.
A. once a day
B. twice a day
C. every two to three hours
D. once a week